Blondes in Venetian Paintings, the Nine-Banded Armadillo, and Other Essays in Biochemistry

Blondes in Venetian Paintings, the Nine-Banded Armadillo, and Other Essays in Biochemistry

Konrad Bloch

Yale University Press

New Haven and London

Published with assistance from the Kingsley Trust Association
Publication Fund established by the Scroll and Key Society of
Yale College.

Designed by Nancy Ovedovitz and set in Times Roman type by
DEKR Corporation, Woburn, Massachusetts. Printed in the United
States of America by Vail-Ballou Press, Binghamton, New York.

Library of Congress Cataloging-in-Publication Data

Bloch, Konrad Emil, 1912–
Blondes in Venetian paintings, the nine-banded armadillo, and
other essays in biochemistry / Konrad Bloch.
p. cm.
Includes bibliographical references and index.
ISBN 0-300-05881-0 (cloth)
0-300-07055-1 (paper)
1. Biochemistry. I. Title.
QP514.2.B597 1994
574.19'2—dc20
94-16906
CIP

A catalogue record for this book is available from the British Library.

The paper in this book meets the guidelines for permanence and
durability of the Committee on Production Guidelines for Book
Longevity of the Council on Library Resources.

10   9   8   7   6   5   4   3   2

To the late Rudolph J. Anderson,
professor of biochemistry at Yale University,
who facilitated my immigration to the United States

Contents

_____

Preface

Choices of ongoing activity are few for a scientist on permanent sabbatical. To write an autobiography did not appeal to me. To be successful, personal memoirs require an introspective mind, literary talent, and an author whose background, as well as scientific contributions, are more than ordinary. I did not feel up to this task. Also, to summarize once again my past research for technical journals I found unappealing. Eventually I decided to write a series of essays, drawing on material tucked away in my mind, incidents, episodes, encounters, chance observations during travel. Conversations with colleagues and students and reading outside my field of specialization contributed much to what I wished to say. Not until I faced finding a substitute for experimental research did I retrieve these reminiscences. This book, a collection of essays, was the result. Although the writings are based largely on personal experiences, I did not intend that they form an autobiography.

Most of the topics are peripheral to mainstream biochemistry, not covered in textbooks or biochemistry courses. All are self-contained and could be published separately. If there is a common theme, it might be an attempt to explain the chemical basis of biological phenomena, Nature's motives in arriving at specific chemical solutions for given biological functions. Biodiversity is emphasized.

Necessarily, I had to resort to chemical language for much of what I wanted to say. I puzzled over finding a suitable level for communicating with readers other than my professional colleagues. At one time, but no longer, articles published in *Scientific American* could have been suitable models. Today they have become too technical for a lay reader lacking a minimum background, especially in chemistry—which unfortunately remains a foreign language for most. Therefore I did my best to keep chemical vocabulary and formulas to a minimum. This minimum I consider essential for an appreciation of certain technical arguments. I trust that in these instances the lay reader will still get an inkling of the line of reasoning, at least the gist of chemical arguments.

As for the period covered, I limit it largely to biochemical research

conducted during the first half of this century, to my generation the Golden
Age of Biochemistry. Today molecular biology dominates all of the life
sciences, solving many of the problems that were beyond the reach of
classical biochemistry. Still, not all of biochemistry deserves to be relegated to history. If my look into the past should acquaint some readers
with subjects not widely known, I will feel amply rewarded for my efforts.

That I ultimately became a biochemist, I cannot attribute to family
background. Among my relatives were lawyers and physicians, but not
scientists or academics. In high school my major interest was in engineering, specifically in metallurgy. A text on the chemistry and physics of
metals and their alloys I found fascinating. Stainless steel, an alloy containing chromium and nickel, had just been invented and there was much
optimism that the time had come for a rational metallurgy, the design of
tailor-made alloys with a variety of desirable properties.

When I enrolled in the metallurgy program at the Technische Hochschule in Munich, the courses offered were uninspiring. The chemistry
courses were a different matter. Hans Fischer's lectures in organic chemistry had a lasting impact on me. His presentation of the subject matter
was masterly and superbly organized, though delivered in a monotone. I
learned from Fischer that playing to the galleries does not guarantee
effective information transfer; it is the course content that counts. At the
end of my sophomore year I knew I had found my field.

My studies in Munich terminated—forcibly—with the equivalent of a
master's degree in chemical engineering, although the only relevant course
I had taken dealt exclusively with the process of brewing beer. I did,
however, receive a thorough training in organic chemistry. As a student,
it was my task to prepare various pyrroles and porphyrins, the building
blocks for the blood pigment hemin. Fischer needed these substances for
the total synthesis of hemin, an achievement that won him a Nobel Prize.
Probably at this time I first became intrigued by biosynthesis—Nature's,
not man's, device for constructing complex organic molecules.

After an interlude of a year in Davos, Switzerland, I managed to make
contact with Rudolph Anderson, professor of biochemistry at Yale University. He was the authority on the lipid chemistry of tubercle bacilli,
the subject assigned to me at the time. On my request, he sent me reprints

of his publications. When I turned to him later for help in my efforts to immigrate to the United States, two letters arrived promptly from Yale University: one from Michael Winternitz, dean of the medical school, informing me of my appointment as an assistant in the department of biochemistry; and a second from Anderson, expressing his regret that no funds were available for a stipend to go with the dean's appointment. I took only the dean's letter to the American consul in Frankfurt. Without delay he issued the immigration visa. Perhaps he was a Yale man.

At Anderson's urging I applied for admission and was accepted as a biochemistry graduate student in the medical school of Columbia University.

My doctoral dissertation, directed by Hans T. Clarke, was still in organic chemistry. At the time, the physician-biochemist Rudolf Schoenheimer was the star of Columbia's biochemistry department. His design and numerous applications of the isotopic tracer technique marked the beginning of a new tool, and a new chapter in biochemistry. Isotopic tracers were the answer to a biochemist's prayer, a direct and unequivocal method for discovering how molecules are made and transformed in biological systems.

When Schoenheimer offered me a fellowship, I accepted eagerly. He converted me into an enthusiastic biochemist. One of the research projects he assigned me evolved into a lifelong interest in the biosynthesis of cholesterol. The problem was challenging, starting, as it were, from ground zero and shown ultimately to require more than thirty consecutive steps. One of the many lessons I learned from Schoenheimer was his custom of assigning his students research problems they could regard as their own. Because of this practice, the morale of his group was exceptionally high.

The reader will find in the various chapters, where appropriate, some further autobiographical details.

I am indebted to numerous colleagues for guiding me to unfamiliar literature sources. Essential information came from Floyd Bloom, Kenneth Carpenter, Avram Goldstein, Guido Guidotti, Kenneth C. Hayes, Kurt Isselbacher, Gilla Kaplan, Donald Kramer, Terisio Pignatti, Arthur Solomon, and Leszek Wolfram. This list is incomplete and I apologize to those whose names have been omitted. My thanks go to John T. Edsall,

Alexander Leaf, and Frank Westheimer for reading various chapters. Members of my family kept me from flights of fancy. The common sense, good judgment, and infinite patience of my wife, Lore, provided invaluable support.

For early encouragement to write, I thank Ernst Mayr. I am most grateful to Howard Boyer for guiding me to a receptive publisher, Yale University Press. Its science editor, Jean Thomson Black, has been an ideal shepherd.

Deborah Fass was most helpful, saving me numerous trips to the various Harvard University libraries. Renate Hellmiss expertly prepared the drawings of chemical structures. E. J. Corey and his associates kindly proofread the manuscript. Last, but certainly not least, Lydia Carmosino and Patricia Gall deserve special credit for transferring my handwritten drafts to the word processor.

The writing of this book was aided by a special grant, No. 870 4030, from the National Science Foundation and by the Chemical Research Fund of Harvard University.

Blondes in Venetian Paintings, the Nine-Banded Armadillo, and Other Essays in Biochemistry

I

Blondes in Venetian
Renaissance Paintings

———

A hypothesis is presented that numerous blonde women

in Italian Renaissance paintings were chemical, not

natural, blondes. Peroxide was unknown at the time.

Chemical mechanisms underlying these early

practices are proposed.*

Ordinarily one tends to associate blonde hair with Nordic features, and if you encounter an example of the trait elsewhere—for example in southern Europe—you suspect the occasional blonde, if indigenous and not a tourist, to be more likely chemical than natural.

In his preface to *The Italian Painters of the Renaissance* (Phaidon Press, 1952, p. ix) Bernard Berenson begins by saying, "Many see paintings without knowing what to look at." Most of the general public surely visit museums simply for pleasure—stimulation of the senses, comparable to the enjoyment of a favorite musical composition. I must confess that until a few years ago I also visited art galleries solely for pleasure. No longer.

*An earlier version of this chapter (in Italian) was published in 1990 in *Eidos, Revista di arti letteratura e musica*, Asolo Arte Publ. 5, 21–25.

I

In response to Berenson, I would now say that I know what to look at. I am viewing paintings with a chemist's eye also.

In the 1970s good fortune took me and my family to Asolo, a small Italian hill town in the North, not far from Venice. It became our favorite vacation spot. Visits to Venetian museums, the Correr and the Academia, and annual pilgrimages to the nearby Villa Barbaro in Maser—one of the Palladio villas still completely preserved, and famed for its Veronese frescoes—remain highlights of these annual sojourns. What intrigued me early on was the abundance of blondes in Venetian Renaissance paintings, notably of Titian, Tintoretto, and Veronese, and of less well known masters also. Were some of these blondes chemical rather than natural?

Before reaching a firm conclusion on this issue (as I will), I want to consider several alternative explanations for the blondes Titian and his contemporaries immortalized. First of all, the numerous invasions of the Italian peninsula from the north throughout history may have left their mark on the Italian gene pool, in which event some of the models the painters chose may have been blondes favored by Nature. Second, one might argue that the chosen color—blonde, brunette, black, or various shades in between—may reflect artistic license, not the actual color of the model's hair. If this was the practice, I have not seen it mentioned; but it is difficult to rule out. With few exceptions, Renaissance painters omit any identification of those who posed for them. Models are known for certain only in commissioned portraits, for example in the Veronese frescoes that depicted members of the Barbaro family in the villa of that name. Yet there are no records of whether or not the noble ladies were natural blondes.

My suspicion that most of the Renaissance blondes in question knew how to bleach their hair evolved slowly but grew stronger, the more paintings I inspected during my search. I list here the ones I chose as supporting evidence, obviously selected to prove my point.

Ceremonial	Carpaccio (1450–1526),	*Santa Ursula*
	Tintoretto (1518–1595),	*Feast of Belshazzar*
	Paris Bordone,	*Marriage of Saint Catherine*
Mythological	Titian (1477–1576),	*Venus and Her Mirror*
		Venus and Lutenist
		Rape of Lucretia

Blondes in Venetian Renaissance Paintings

	Veronese (1528–1588),	*Venus and Adonis*
		Venus and Mars
		Lucretia
Frescoes in Veneto Villas	Veronese, Villa Barbaro,	*Lady with Lute*
		Retreat of Elena Capliari
	Giovanni Antonio Fasolo (16th C),	*Concert*
		Cellist
Portraits	Veronese, Tintoretto,	*Portrait of a Woman*
		Portrait of Bella Nanna
		Lady in Red
Florence	Botticelli,	*Primavera*
		Portrait of a Lady

The paintings tend to be secular, portraying themes popular during the Renaissance, allegorical or taken from mythology, or occasionally biblical. More often than not, however, the madonna's hair, if visible at all and not covered, tends to be brunette or even darker. One notable exception— and there are others—is Raphael's blonde *Little Cowper Madonna* in the National Gallery in Washington, D.C.

More common than rare, the blondes are luxuriously dressed (if dressed at all). Strings of pearls adorn their necks and coiffures, denoting wealth either because they belong to the upper classes or because they are courtesans, distinctions that might be blurred, cutting across the lines of respectability. One further argument for my hypothesis seems to me compelling. If males of the species accompany the blondes in the same paintings, their hair color is generally dark. Veronese's *Venus and Adonis* and *Venus and Mars*, or Titian's *Venus and Lutenist* and Tintoretto's *Feast of Belshazzar* are striking examples. Furthermore, so far I have not come across any examples of blue-eyed blondes, a trait that more often than not marks a Nordic type. Certainly exceptions to my generalizations exist, but by themselves they do not invalidate the hypothesis.

Eventually my suspicion that some of the blondes portrayed in numerous Venetian Renaissance paintings were chemical and not a caprice of Nature grew into firm conviction. Persuasive support came from two sources in the literature, thanks to the help of Terisio Pignatti of the University of Veni~e. The treatises he recommended were *Les Femmes Blondes Selon les peintres de l'école de Venise* by one Armand Brachet, published in 1865, and an Italian account, *Curiosità Veneziane,* by Giuseppe Tassini, dated 1887. Both chronicles address at length the question of whether the Venetian blondes are whims of Nature or miracles performed by the pharmacist. In page after page they describe the fashions and notions popular among sixteenth-century Venetian nobility, especially of the female gender. The writers highlight the desire of these women to attain or emulate the idealized images of Grecian antiquity, the fabled beauty of Venus, Ceres, Psyche, or the Three Graces, all blondes according to the Renaissance literature. Of course, records of Greek paintings do not exist except as vase decorations.

Tassini tells us in great detail how the noble Venetian ladies in Renaissance times went about achieving this ideal of turning from brunette to blonde. To quote him: "Where the sun is strongest they spend hours on the balconies of their mansions or palazzi, bathing and rinsing their hair with a tincture known as 'aqua bionda,' or 'aqua di gioventu,' then drying the tresses and repeating the procedure several times [Fig. 1.1]. They cover their shoulders with a silken shawl or kerchief, named schiavonetta. On their heads they wear a straw hat, called 'solana,' for shading their faces, which has a circular opening at the top leaving their hair exposed to the midday sun." Tassini's account deals only with the manipulations in use, the extended and repeated exposure to the sun. He does not comment on the chemistry of aqua bionda.

Venetian ladies were not the only ones who changed the color of their hair. A popular jingle of the time ran:

Florentine women bleaching their hair—
for their households they have no time to spare

The striking blondes in Botticelli's *Primavera* (flanked on the left by a dark-haired youth) were painted before 1500. Some sort of aqua bionda thus was certainly known to Florentines as well and may have been discovered independently in various Italian regions. Some examples follow.

In about 1500 Caterina Sforza published her *Experimenti,* describing

Fig. 1.1 Donna veneziana sull'altana con la solana. (From Cesare Vecellio, *Degli habiti antichi e moderni di diverse parti del mondo*, Venice, 1590.)

her recipes for bleaching hair. Lucrezia Borgia, traveling to meet her third husband in Ferrara, stopped several times en route to restore her hair to the blonde shade she desired. Even earlier dates for the invention of aqua bionda can be deduced from a recent review of a Metropolitan Museum of Art exhibit entitled "Paintings in Renaissance Siena" (Manuela Hoelterhoff, 1989). San Bernadino, a Sienese saint, is said to have upbraided the women of Siena for "bleaching their hair in the sun, washing and drying, washing and drying . . . "

Before turning to some chemical thoughts on the subject, I will relate the story of Caterina Cornaro, to this day a local heroine of the Asolo populace. Born in Venice and raised in a Padovan convent, she became the spouse of the King of Cyprus. I first came across an etching of Regina Cornaro in Asolo's Civic Museum and later learned that Titian had painted her portrait. Alas, only a copy, not the original, is on display in the Uffizi Gallery in Florence. A local historian, Juliette Benzoni (Asolo, 1983 p.3), describes her vividly as follows: "Her blondeness [*biondezza*] was that well known to Venetian women and—as legend has it—stolen from the sun. As for her eyes, they were dark as the night [*un nero d'inferno*], the largest and sweetest of the world." Among my examples Caterina is the only one displaying so clearly and unequivocally the combination of blonde hair and pitch-black eyes.

Readers interested only in the cultural or aesthetic phenomenon of Venetian blondes may well wish to put this chapter aside here. They should be aware, however, of the historically important fact, known perhaps only to chemists, that hydrogen peroxide was not discovered until early in the nineteenth century! At any rate, to the inquisitive chemist it would be an interesting challenge to identify the ingredients of aqua bionda and to reproduce the rinses that are said to have turned brunettes into blondes.

Chemistry of Aqua Bionda

In today's culture we speak of peroxide blondes, in distinction to natural blondes. Hydrogen peroxide (H_2O_2), the common bleaching agent, does the trick and to the best of my limited knowledge of the subject no substitutes for peroxide have so far been invented. A recent patent claims success in bleaching hair by laser radiation, but it seems doubtful that the Venetian ladies were aware of that procedure. Hydrogen peroxide was discovered by Thenard as late as 1812. He prepared it by heating barium peroxide with acid. It was first advertised in the 1870s as a hair bleach by the French pharmacist Thielly under the label "Eau de Jouvence." Did the rinses, lotions, or concoctions used some four hundred years ago, or perhaps even earlier, in fact contain hydrogen peroxide?

Tantalizing clues to this effect can be found in the lengthy appendix to Armand Brachet's treatise. Listed here are no fewer than thirty-six different recipes for preparing aqua bionda. According to Brachet's "Recette per Biondare," every one of the prescriptions claiming to turn brunettes into blondes specifies aqueous extracts of various higher plants as starting materials:

Plants Extracted
Licorice (glycerin)
Boxwood[a]
Flowers of walnut
Myrtle leaves
Cypress blossoms[a]
Cumin (herbal medicine)[a]
Lupin[a]
Myrrh (sweet cicely)[a]
Dregs of white grapes[a]
Buds of poplar
Madder root[a] (contains alizarin,[b]
 orange-yellow red)
Clover[a]
Holly
Aloe
Ashes of grape vines

Chemicals
Saltpeter[a]
Powdered silver
Quicklime
Alum[ac]
Wood ash (potassium carbonate)

a. mentioned in several recipes
b. anthracene derivative
c. used for bleaching raw silk

The procedures for preparing aqua bionda are as follows:

Step 1 Leaves and (sometimes) green stems extracted with water; extracts placed in shallow trays, and exposed to sunlight for one to two weeks.

Step 2 a. *Concentration* of an active bleaching reagent already present, *or*

b. *Production* of an active bleaching reagent.

Step 3 Hair rinsed repeatedly in strong sunlight.

Sunlight and air (oxygen) are needed in steps 2 (either a or b) and 3.

Potential Mechanisms Some plant substances oxidized by air oxygen in presence of strong sunlight yield either hydrogen peroxide or other "active" forms of oxygen. Obviously, the exposure to sun will evaporate much of the water, and this raises the first question of chemical interest. Does the process of aeration and exposure to sunlight simply serve to concentrate extracts already containing the putative bleaching agent, or does the treatment produce it? In the absence of appropriate control experiments one cannot give an answer. What follows is an exercise, entirely speculative, to prepare the ground for laboratory work that someone so inclined might wish to undertake for explaining the bleaching potency of the Venetian aqua bionda.

Melanins, the pigments of mammalian hair, are complex molecules derived oxidatively from the amino acid tyrosine—pigments entirely different chemically from the wider range of colors displayed by the plumage of birds. Melanins are of two types, eumelanins (black) and pheomelanins (red) occurring apparently side by side. In albinos both pigments are absent, resulting from a genetic deficiency of the tyrosine-converting enzyme(s) known as tyrosinases. The different shades of hair color are due to varying proportions of eumelanins and pheomelanins. Extensive bleaching experiments with female human hair have been carried out by Wolfram and Albrecht (1987) at the Clairol Research Laboratories. Their principal results show that both solar radiation and oxygen are needed for bleaching, brown hair lightening more readily then red and becoming more reddish in the process. These authors formulate the bleaching process as

$$2 H_2O + 2O_2^- \rightarrow H_2O_2$$

where O_2^- is an "active" form of oxygen produced by solar radiation of moist air. This species subsequently dismutates to H_2O_2, the active ingredient of all commercial hair bleaches. So far the ancient procedures for preparing aqua bionda are in harmony with modern chemistry.

Knowing the reagents and conditions required for bleaching hair and the role of the two melanins (eu- and pheo-), we are left with the crux of our problem: what is the chemical in plant extracts that furnishes hydrogen peroxide or its equivalent?

One candidate molecule is ascaridole* (Fig. 1.2), found in a few plants of the goosefoot family (chenopodium), which includes both herbs and woody plants. Ascaridole is structurally classified as a transannular terpene endoperoxide. Solar radiation of α-terpinene in the presence of chlorophyll produces ascaridole also in the test tube. The α-terpinene itself is ubiquitous in higher plants, richest in marjoram, coriander, mosla, and cardamon. If ascaridole is capable of yielding H_2O_2 under the conditions specified for producing aqua bionda, one of our major questions would be answered. Chemical arguments pro and con are suggestive but not compelling. Nor is the possibility eliminated that ascaridole as such is a bleach.

Ascaridole or a related endoperoxide remain strong contenders because another family of plant constituents, derivatives of anthracene (Fig. 1.2), form endoperoxides, which in turn are capable of producing H_2O_2. The precise structure of this tricyclic endoperoxide can only be inferred from the fact that the anthracene derivative alizarin, the reddish-orange dye from Rubia tinctorum known since antiquity, is a plant product.

A list of reactions or sources claimed to produce or contain H_2O_2 ever since its discovery would include rainwater after a thunderstorm, the irradiation of algae or juniper, and the autoxidation of turpentine. Whether these possibilities are real or not, the concentration of H_2O_2 so produced would undoubtedly be too small to be effective.

One other more probable and well-documented substantial source of H_2O_2 deserves mention. Solutions of ascorbic acid (vitamin C), abundant not only in citrus fruit but in all plants, yield H_2O_2 on aeration in the

---

*The name derives from its antihelminthic properties of protecting vertebrate hosts against parasitic nematodes such as Ascaris.

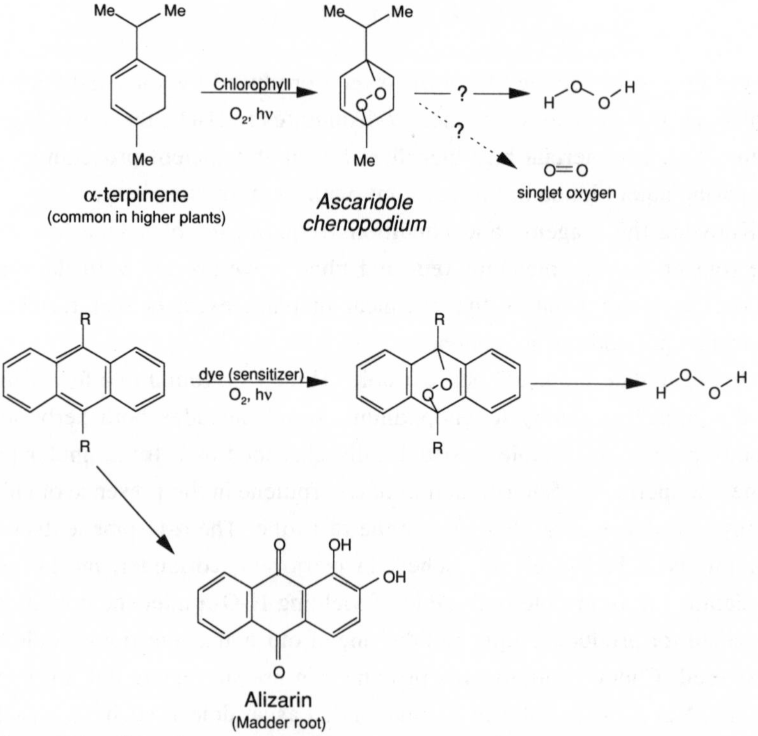
Me
Me
Chlorophyll
O_2, hv
Me
α-terpinene
(common in higher plants)
Me
Me
Me
Ascaridole
chenopodium
?
?
singlet oxygen
R
R
dye (sensitizer)
O_2, hv
R
R
OH
OH
Alizarin
(Madder root)

Fig. 1.2 Possible plant sources of hydrogen peroxide from terpenes and anthracene via endoperoxides.

presence of Cu^{++} (Fig. 1.3). But lemon juice is not mentioned in any of the Venetian recipes. In any event, the vitamin alone would only lighten, not bleach.

Air or oxygen in conjunction with the radiant energy of sunlight destroys color very effectively. Kodachromes or lithographs fade unless protected by screens that filter out ultraviolet light. According to current notions, the molecular species that causes certain colors to fade is singlet oxygen, an active oxygen species that arises photochemically from ground-state, or triplet, oxygen. Passing ground-state oxygen through a solution containing a so-called photosensitizer, either a synthetic dyestuff or chlorophyll (the universal green plant pigment), generates the energy-rich singlet oxygen.*

*The energetically lowest electronic configuration of oxygen, the ground

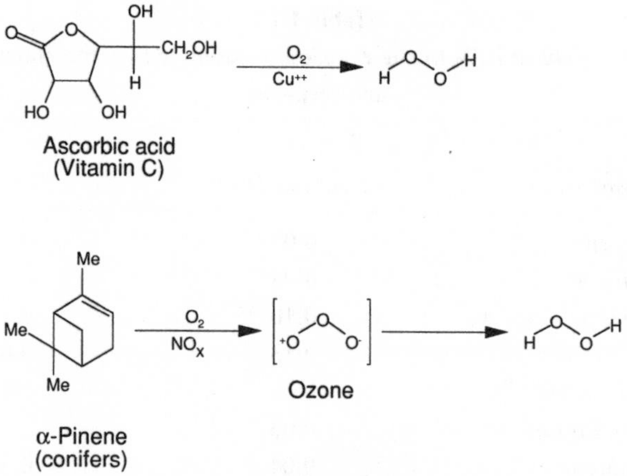

Fig. 1.3 Potential conversion of ascorbic acid and
α-pinene to hydrogen peroxide.

Thérèse Wilson (Department of Biology, Harvard University) raised the possibility that during the preparation of aqua bionda singlet oxygen might be produced and account for its bleaching properties. However, experiments in her laboratory with snippets of student-donated black hair failed. This failure should not have come as a surprise. I did not learn until afterward that singlet oxygen attacks mainly organic molecules containing double (or olefinic) bonds, structural features absent in melanins.

Finally, I must mention some recent experiments that implicate ozone as a potential oxidant for producing H_2O_2 by reaction with plant terpenes. Becker and colleagues (1990) point out that moist forest air, especially above decaying leaves, contains significant amounts of ozone. In their test tube experiments (Table 1.1), several plant terpenes exposed to ozone produced H_2O_2. Limonene, a terpene from citrus fruit, yielded as much as 1.8 percent hydrogen peroxide (commercial hair bleaches contain about 3 percent H_2O_2). Whether their in vitro conditions are realistic remains to be proven. Still, as mentioned earlier, the Venetian recipes call for lengthy exposure of plant extracts to the sun, conditions that would substantially raise the H_2O_2 concentration of the initial extracts by evaporation.

---

state, is known as triplet oxygen. Singlet oxygen $^1\Delta gO_2$, an electronically excited state, is produced by exposure to photosensitizers such as dyestuffs or chlorophyll.

Table 1.1

Molar yield of H_2O_2 in the reaction of ozone with some alkenes and terpenes

Hydrocarbon	% H_2O_2 (without H_2O)	% H_2O_2
Isoprene	0.036	0.1
β-Pinene	0.04	0.15
α-Pinene	0.10	0.5
Δ^3-Carene	0.13	0.6
D-Limonene	0.30	1.8
trans-Butene	0.06	0.5
Isobutene	0.04	0.37

Source: Becker et al., 1990.

My summary conclusion is that the aqua bionda prepared in the fifteenth and sixteenth centuries could well have been potent bleaches, with H_2O_2 as the most likely active ingredient. There is a certain rationale to the Venetian recipes; they cannot be discredited as alchemy.

History of Bleaching Hair

While this essay focuses on the blondes in Italian Renaissance paintings of the north because of the personal interests I mentioned, bleaching of hair was apparently practiced very much earlier (Wall, 1962). According to this author, Roman ladies greatly admired the golden hair of the captives brought from northern countries and tried to imitate it, or demanded that the hair be shorn from captives and made into wigs for their use. References to these practices can be found in writings as early as those of Pliny. "Blond washes" used during Roman times already contained many of the mineral ingredients (such as alum, wood ash, and quicklime) cited in Brachet's "Recette per Biondare," but some of these early recipes oddly included goat's fat. Because animal fat when treated with potash or quicklime produces soap, Wall makes the intriguing suggestion that the popular soap, imported from Gaul or Germany, was originally invented for bleaching hair, not for cleansing. Whatever the preparation of aqua bionda, the use of common plant concoctions must probably be dated to the Renaissance period. Wall, recording bleaching practices throughout history, makes another intriguing suggestion about their history. As noted earlier, following Thenard's discovery of hydrogen peroxide in 1812, the oxidant was first commercialized during or after the Paris Exposition in 1867 as "Eau de Fontaine de Jouvence Golden," a few decades after the custom of wearing false hair, peruques, or wigs during the seventeenth and early eighteenth centuries became unfashionable. Revival of hair bleaching was therefore an instant success. The commercial "Better Living through Chemistry," unfortunately no longer used today, would have been appropriate.

Bibliography

1. C. Vecellio (1590), *Degli habiti antichi e moderni di diverse parti del mondo*. Venice.

2. A. Brachet (1865), *Les femmes blondes selon les peintres de l'école de Venise*.

3. G. Tassini (1887), *Curiosità veneziane*, ed Filippi, pp. 541–542. Venice.

4. E. Rodocanachi (1907), *La femme italienne a l'époque de la Renaissance*, p. 111. Paris, Librairie Hachette.

5. D. Reato (1988), *Le maschere veneziani*, ed. Arsenale. Venice.

6. M. Hoelterhoff (1989), Paintings in Renaissance Siena, *Wall Street Journal*, March 9.

7. F. E. Wall (1962), *Principles and practice of beauty culture*, 4th ed. New York, Keystone Publications.

8. L. J. Wolfram and L. Albrecht (1987), Chemical and photobleaching of brown and red hair, *J. Soc. Cosmet. Chem.* 82, 179.

9. K. H. Becker, K. J. Brockman, and J. Bechara (1990), Production of hydrogen peroxide in forest air by reaction of ozone with terpenes, *Nature* 346, 246–250.

2

Evolutionary Perfection of a Small Molecule

———

In the words of Aristotle's *Politics,* "Nature is the end,

and what each thing is when fully developed we call

'Nature.'" Not only genes but also small molecules

changed in the course of evolution. The example cited

here is the evolving structure of the sterol molecule.

Along with structural changes, the functions of sterols

improved and diversified. It is proposed that increasing

atmospheric oxygen concentrations were the driving

force behind these changes.

The first research assigned to me during my stay in Davos, Switzerland, in the mid-1930s may have planted the seeds of my lasting interest in the subject of this chapter. I was to resolve a controversy about whether or not tubercle bacilli contained cholesterol. My predecessor at the Schweizerisches Institut für Höhenforschung had obtained positive results, with which Erwin Chargaff, then at Yale University, disagreed. Who was right? At the time, in 1934, the standard assay method called for isolation of cholesterol as the insoluble digitonide, a procedure my predecessor had used. I confirmed his findings—up to a point. In my hands, the material yielding the insoluble "sterol" digitonide was not cholesterol but an aliphatic hydrocarbon produced by certain bacteria, among them tubercle bacilli. Whether or not bacteria (prokaryotes) contain cholesterol as do all animal species and yeast (eukaryotes) had not yet been investigated systematically. Any generalizations about this issue would have been premature. Still, the results obtained with tubercle bacilli, recorded in my first published paper,* may have stimulated my developing interest in comparative biochemistry, a common theme of several of these chapters.

For several generations biochemists have embraced the credo of biochemical unity (or universality) at the molecular level, and with considerable justification. Major metabolic pathways and certain modes of energy production have turned out to be identical in all extant forms of life. Biochemical unity has in fact become one of the pillars supporting Darwin's theory of common descent. It is overwhelmingly valid for the genetic code (the invariant structures of the twenty amino acids that proteins contain) and, with important exceptions, for metabolic pathways. Minimal processes essential to all living cells are indeed universal, remaining apparently unchanged during evolution from the most primitive cells to man.** In some respects, however, the universality concept needs to be qualified.

As I will discuss in the next chapter, the advent of atmospheric oxygen has wrought fundamental changes in cell morphology, modes of energy production, and metabolic control mechanisms. At the molecular level, Nature's invention of the sterol molecule, once environments became aerobic, exemplifies these fundamental innovations. First of all, choles-

---

*Of dubious relevance here, but important to me at the time, papers published in Hoppe-Seyler's *Zeitschrift für Physiologische Chemie* before World War II carried a substantial remuneration for the author. Today the cost of publishing in all scientific journals and of reprints from them, takes a major bite out of research budgets.

**The existence of an "RNA world," one without DNA, seems increasingly possible but remains a hypothesis. Also, whether life originated from inanimate matter more than once is a philosophical question.

terol's functions are unusually diverse. Second, and equally important to me, a stepwise evolution of this molecule's structure and function proved exceptionally amenable to experimental analysis. Evidence that organismic evolution and the evolution of a small molecule progressed side by side will be presented in what follows.

Oxygen and Sterol Biosynthesis

A few years after my experience abroad with tubercle bacilli, my postdoctoral supervisor and mentor at Columbia University, Rudolf Schoenheimer, came up with an idea, brilliant in retrospect. He had been long interested in the biological synthesis of cholesterol. Was the single oxygen atom in cholesterol derived from water as the oxygen donor in an anaerobic process, or was atmospheric oxygen the source? Up to that time oxygen was thought to serve exclusively as the terminal electron acceptor in respiration, but not as a source of carbon-bound oxygen in organic compounds. I began my assignment with enthusiasm but for technical reasons was unable to solve the problem. Years later (in 1957), after I had joined the Harvard chemistry department, my associate T. T. Tchen* proved that indeed molecular oxygen was the donor for cholesterol's OH group. Clearly, cholesterol is not a universal molecule but is produced only after the advent of aerobic metabolism. In fact, at least six of the numerous steps in cholesterol biosynthesis require molecular oxygen.

In this chapter the emphasis will be on the terminal stages of sterol biosynthesis, the steps that streamlined an "ursterol" and in the process improved and perhaps extended and perfected function, including the precursor role of cholesterol for numerous steroid hormones. In essence, our experimental results suggest a replay of evolution—natural selection based on chemical rationale.

To set the stage for the argument, we cannot avoid outlining the oxygen-

---

*To the best of my knowledge, the question Schoenheimer posed remained dormant until, fifteen years later, L. Ruzicka and colleagues at the Eingenössische Technische Hochschule in Zurich, proposed that OH$^+$, an electrophilic species of oxygen, initiates the cyclization of squalene.

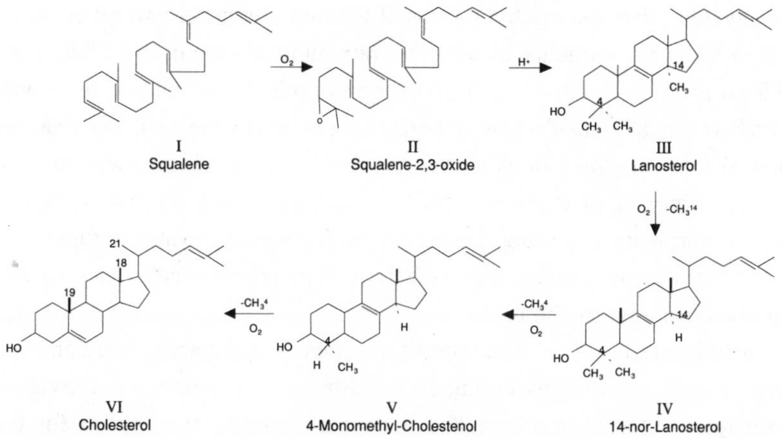

Fig. 2.1 Cyclization of squalene-derived-2,3-oxide to lanosterol, and sequential demethylation of lanosterol to cholesterol. I, squalene; II, squalene oxide; III, lanosterol; IV, 14-nor-lanosterol or 4,4-dimethyl cholestenol; V, lophenol or 4-monomethyl cholestenol; VI, cholesterol. In VI, R=H; in the plant sterols stigmasterol and β-sitosterol, R=C_2H_5.

requiring chemical events in cholesterol biosynthesis (Fig. 2.1). Perhaps even those readers unfamiliar with the language of chemistry will perceive the gist.

On the Importance of Being Demethylated

An oxidative attack on the hydrocarbon squalene to form squalene oxide (Fig. 2.1, I → II) initiates the cyclization of this molecule to lanosterol* (II → III), the first tetracyclic intermediate on the way to cholesterol. In 1966 this was shown independently by E. J. Corey and E. Van Tamelen. Further transformations involve, in the main, the sequential removal of three "extra" methyl (CH_3) groups that protude from the otherwise planar

*Lanosterol is so named because in the animal body the sterol occurs in substantial amounts in wool fat (lanolin) but only in traces in internal organs or tissues. It has no biological function except as a short-lived intermediate on the way to cholesterol.

ring system, first the methyl groups at C_{14} and thereafter two others at C_4 (III → VI). This sequence of methyl-group removals seems to be invariant, at least in animals and yeast. Two aspects of this demethylation sequence, which occur at considerable expense of energy to the cell, became our central interest. Why does demethylation occur at all, and why are the groups removed in a given order? Asking why—and not how—implied that a purpose, a driving force, or, in Darwinian terms, a functional advantage accrues from natural selection. A priori chemical reasoning was not likely to answer the query.

To address the "why" issue, two experimental approaches were chosen. One, widely used for assessing sterol "fitness" for membranes, involves insertion of sterols into so-called liposomes, models substituting for the natural membranes where cholesterol normally resides. Structural details of the inserted sterol molecule will cause either an increase or a decrease of a chosen physical parameter such as membrane fluidity (or its inverse, viscosity; see Fig. 2.2A). Increase of the "order parameter" signifies a more ordered membrane structure, and decrease indicates less organization or, in extreme situations, chaos. Cholesterol-containing liposomes display high concentration-dependent viscosity values, not exceeded by any known sterol, whether naturally occurring or man-made. When lanosterol, the precursor containing the three "extra" methyl groups, replaces cholesterol in liposomes, viscosity rises only marginally. However, the intermediates in the sequential biological demethylation of lanosterol to cholesterol progressively raise the membrane viscosity and, most significantly, they do so in precisely the same temporal order as the cell's enzymes remove the "extra" methyl groups: 14-methyl → 4′4-dimethyl → 4-monomethyl → cholesterol (Figs. 2.1 and 2.2A). Results obtained with artificial membranes are technically simple but of unproven relevance to biological phenomena, for the reason that liposome models do not contain proteins, the invariant natural components of membranous cell envelopes.

For studying the fitness of the intermediates mentioned in living cells, bacteria known as mycoplasmas or mollicutes turned out to be eminently useful. Uniquely among procaryotes, mycoplasma cells require cholesterol for growth. Notorious parasites, they invade and infect cholesterol-rich animal tissues. In the test tube the bacterial growth rate is proportional

Evolutionary Perfection of a Molecule

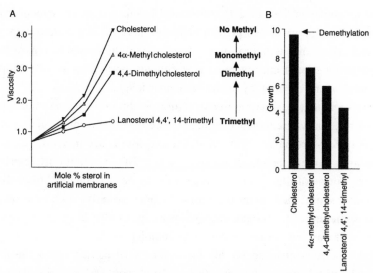

Fig 2.2 A. Effect of increasing sterol concentrations on viscosity of artificial membranes (liposomes). B. Effect of sterols on comparative growth rates of *Mycoplasma capricolum*.

to the cholesterol supplied in the culture medium.* Cultured mycoplasma cells respond to lanosterol only sluggishly (Fig. 2.2B). As noted, this early cholesterol precursor is also ineffective as a modulator of membrane viscosity in artificial membranes. Proceeding along the demethylation pathway, 4.4′-dimethylcholesterol, the next intermediate, promotes significantly faster growth than lanosterol, and finally, removal of one of the two remaining methyl groups at C_4 improves bacterial growth even further— almost, but not quite, matching the growth rate that cholesterol itself affords. Thus we observed a striking improvement of function in vivo as a result of successive streamlining of the lanosterol molecule, parallelling the physical responses elicited by the various sterol intermediates in model membranes.

It is surely no coincidence that the events recorded in Fig. 2.2 follow the same order that eukaryotic cells employ today for converting lanos-

*I owe this information to Shlomo Rottem of Hebrew University.

terol to cholesterol (Fig. 2.1). Most significant, the sterol's membrane fitness and support of cell growth approach perfection gradually and, as we will argue, in a temporal order. It is certainly tempting to regard this stepwise functional improvement and ultimate perfection as an example of evolutionary chronology at the molecular level.

Any argument postulating the structural evolution of a small molecule would be greatly strengthened if there existed cellular or molecular equivalents of fossil remains.* Yet, by definition, fossils are mineral, petrified structures, largely devoid of organic matter. They contain only imprints of early forms of life. There are organisms to be regarded as "fossils" in a broader sense, still extant but probably more primitive on a time scale. Thus some ancient, still-existing organism appears to have invented the sterol pathway but without taking it to completion.

Only one organism so far fits the category of having taken the first innovative and productive step, the oxidative conversion of squalene to lanosterol. The nematode *Panagrella redivivus,* an invertebrate belonging to the free-living flatworms (helminths), produces lanosterol, accumulating but not converting it farther along the sterol pathway (Fig. 2.3).

A second example, the bacterium *Methylococcus capsulatus,* is especially intriguing. Growing on methane (CH_4) as the sole carbon source, this bacterium is necessarily aerobic, equipped with oxygenase enzymes for converting the hydrocarbon into biomolecules. Perhaps for this reason the bacterium has also learned to deal more generally with CH_3 groups. *Methylococcus* does indeed produce and metabolize lanosterol further, converting it to 4,4-dimethylcholesterol and smaller amounts of 4-monomethyl sterol, eliminating two of the three "extra" methyl carbons. Yet the bacterium stops short of completing the pathway. Of special interest to the cognoscenti, *methylococcus* oxidizes the C_{14} sterol methyl groups only partially to formic acid, HCOOH, while the remaining methyl groups at C_4 are oxidized all the way to CO_2.

Finally, organisms exist that accumulate 4-monomethyl sterols, the ultimate precursors of cholesterol. They are the dinoflagellates, planktonic

---

*It has been estimated that 99 percent of all organisms that ever existed are now extinct. Is it not possible that some biomolecules have suffered the same fate, that they are simply no longer around?

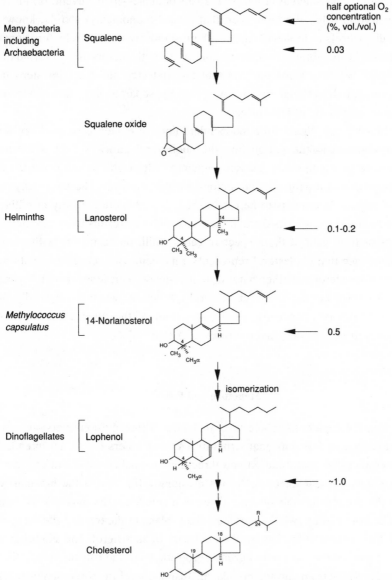

Fig 2.3 Organismic chronology of the sterol structure. The values given for oxygen concentrations (listed in the column to the right of the chemical structures) will be explained later.

organisms classified as either protozoans or algae—in any event, primitive eukaryotic cells. We have seen that their 4-monomethyl sterols, known as dinosterols or lophenols approach most closely cholesterol's functional fitness as membrane reinforcers and as growth factors for sterol auxotrophs. We may also conclude that the first step and the later steps in lanosterol demethylation are catalyzed by separate enzymes, some of earlier and some of later origin.

In summary, then, three independent lines of evidence support—or do not contradict—the notion that the cholesterol pathway from squalene evolved in a temporally discrete sequence. Although one can construct a chemical and organismic chronology for this pathway, the phylogeny of the organisms remains to be established. Nor is there any way of telling whether a given intermediate in the sterol pathway results from mutational loss or from gain of the respective genes. Still, one can state with some confidence that cholesterol represents a molecule perfected by evolution. No other sterol, whether natural or man-made, surpasses it with regard to what we believe to be cholesterol's function, whether physically in membranes or chemically, as the precursor of the diverse steroid hormones in vertebrates, invertebrates, and plants.

Structure and Function

Fig. 2.4 depicts the cholesterol molecule in four different versions, the traditional two-dimensional structure (A) and others (B, C, D) intended to convey the conformation and three-dimensional arrangement of atoms in space. Critical for cholesterol's membrane function is the biplanarity of the tetracyclic ring system. Thus, in a side view the molecule is seen to be bounded by two flat surfaces, the β-plane at the top and the α-plane at the bottom (Fig. 2.4D). By contrast, in lanosterol, the cholesterol precursor, the extra methyl groups protrude from the α-face, humps that render the bottom surface nonplanar. Inspection of the three-dimensional structures therefore allows us to rationalize the surgical elimination of the methyl groups. Models based on physical data (x-ray diffraction patterns) visualize cholesterol's entry into membranes intercalating readily between

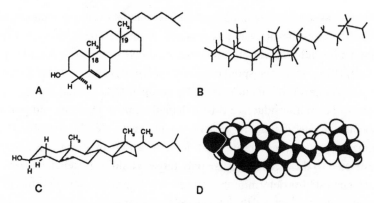

Fig. 2.4 Progressive refinement of the cholesterol structure. A, planar model
(1937); B, Dreiding model (1953); C, conformation model (1956);
D, Pauling-Corey space-filling model (1960?).

the linear fatty acid chains of the phospholipids, the membrane's matrix
(Fig. 2.5). Physiological fitness, in this case membrane stabilization, results from intimate physical contact between lipids, which cholesterol
provides but not lanosterol.

As described, streamlining of cholesterol precursors by expulsion of
methyl groups occurs at the α-face, the underside of the sterol molecule.
Methyl substituents elsewhere, for example those attached to C_{18} and C_{19}
(known as angular methyl groups; see Fig. 2.4A), cells leave in place, with

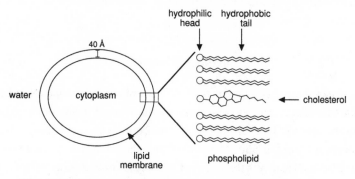

Fig. 2.5 Schematic presentation of cholesterol residing in the
phospholipid membrane bilayer.

one known exception. One finds such an exceptional event upon chemical analysis of marine sterols, such as sponges or corals. These invertebrates elaborate a bewildering variety of exotic, modified sterol structures (Djerassi et al., 1979). Thus, the sponge *Axinella polypodes* converts cholesterol to a nor-cholesterol, eliminating the CH_3 group at C_{19}, in what one would presume to be an unproductive β-face demethylation. That this additional step occurs at all, if only in an isolated instance, clearly weakens the generalization attempted above, that demethylations of sterol precursors are restricted to the α-face because only those would be beneficial events. In fact, our tests for determining sterol fitness (Fig. 2.2) show that the 19-nor sponge sterol is markedly inferior to cholesterol.

Why this unproductive, even deleterious, step? It is a dilemma, but one can escape it by adopting François Jacob's famous concept of "evolutionary tinkering," Nature's operating by trial and error. Can one further argue that the isolated case of sterol β-face elimination practiced by some sponges was abandoned during the evolution of organisms because it was a retrograde step? Sponges indeed appear to be on a dead-end branch of the evolutionary tree. Today the available techniques of gene transfer from one organism to another might disclose whether this phenomenon is an example of retrograde evolution. The cloned sponge gene responsible for enzymes that produce the 19-nor sterol, transferred to some other cholesterol-producing organisms, if expressed may or may not have deleterious effects on the recipient cell's viability (for instance the integrity of membranes or the production of steroid hormones).

Sterols and Insects

The insect world provides us with an example of what appears to be an evolutionary adaptation evident from the nutritional needs of certain invertebrate species. Insects require sterols from an external source during development from the pupae to the larval stage as shown by G. Fraenkel (1941). Cholesterol meets the needs of all insects studied. Like vertebrates, however, some invertebrate species practice a herbivorous lifestyle while others are carnivorous. Meat eaters have access to an ample dietary supply of cholesterol; herbivores do not, because the leaf sterols of higher plants contain two extra carbon atoms (C_2H_5) in the aliphatic side chain. For all insects studied, plant sterols are useless as such because they are not absorbed from the gut into the circulating hemolymph.* Plant eaters such as the silk worm *Bombyx mory,* feeding preferentially on leaves of the mulberry tree, have solved this problem by acquisition of enzymes that remove the interfering C_2H_5 group and conversion of plant sterols to cholesterol. On the other hand, the omnivorous cockroach, when foraging on animal matter containing cholesterol, need not take the trouble of modifying the plant's sterol sidechain. Metabolic and nutritional versatility may explain the extraordinary reproductive capacity of this vermin. By the same token, for strictly carnivorous insects such as the hide beetle *Dermestes vulpinus,* the enzymes converting plant sterol to animal sterol should be dispensable.

For reasons we do not yet know, the brush-border region of the human intestinal tract is also impermeable to plant sterols; they pass unchanged into the larger bowel and feces. Only in a few instances of a metabolic disorder known as phytosterolemia (altogether a total of twenty-two cases has been reported) do plant sterols enter the general circulation. The afflicted tend to develop atherosclerosis. But the normal human, even the strict vegetarian, need not worry. Not only are the sterols of plant origin innocuous but they may even be beneficial, because they antagonize and therefore lower the intestinal absorption of dietary cholesterol. Before the recent advent of effective cholesterol-lowering drugs, plant sterols were prescribed as hypocholesteremic agents. Attractive and popular medications, they are cheap natural products, not designed in the chemical laboratory and therefore not subject to approval of the Food and Drug Administration. Yet the regimen is not palatable: to be effective, plant sterols must be taken in large quantitites of 10–15 grams per day.

---

*If memory from my youth serves me correctly, a chemical called Eulan (3β-chlorocholestanol) was widely used in Europe as a mothproofing agent. Although it was not known at the time, Eulan interferes with the absorption of cholesterol from the insect's gut. The chemical was accidentally discovered, not rationally designed.

Evolutionary Perfection of a Molecule

Cholesterol-Derived Molecules

Students of comparative biochemistry will note that the further metabolism of cholesterol to specialized molecules is extraordinarily diverse even in the same kingdom. The class of bile acids is an example. In all mammals the major route for disposing of excess cholesterol is oxidative, hepatic removal of the terminal three carbon atoms of the sterol side chain to form cholic acid. This C_{24} carboxylic acid occurs as a major constituent of animal bile (Fig. 2.6A) Also, three hydroxyl groups are introduced into the ring system but differ in position and sterochemistry depending on the species. Their prefixes—cheno- (goose), urso- (bear), hyo- (hog), and pytho- (snake)—are indicative of this bewildering and unexplained species diversity. Since bile acids are secreted into the small intestine and subsequently reabsorbed (enterohepatic circulation), compositional differences in the intestinal flora may be responsible for the structural diversity.* Common to these various bile acids is their sharing of the ability to disperse water-insoluble fat molecules preparatory to absorption from the gut. They are excellent emulsifiers or detergents.**

Of a different kind are the bile acids of the Greenland shark (*Scymnus borealis*) and the related ray. These species convert cholesterol to scymnol, a twenty-seven-carbon alcohol leaving cholesterol's carbon skeleton intact (Fig. 2.6A). Scymnol is not an emulsifier but acquires this property by reaction with sulfuric acid, forming what is known as an anionic detergent, a "primitive" molecule confined to and probably typical of the elasmobranch sharks and rays (cartilaginous fish). This probably ancient vertebrate group may not yet have invented the oxidative conversion of cholesterol to bile acids. The coalecanth, paradigm of extant ancient fish, living in the depth of the Indian Ocean, also produces the scymnol structure.

*Bile acids of identical structure are formed in liver tissue of the rat, the mouse, the rabbit, and man. This fact probably indicates no more than a similarity of the bacterial flora residing in these species.

**The mammalian detergents are not the free bile acids but "bile salts," conjugates with taurine in carnivores, with glycine in herbivores, and with either in omnivores. The two conjugate types appear to be equally effective detergents.

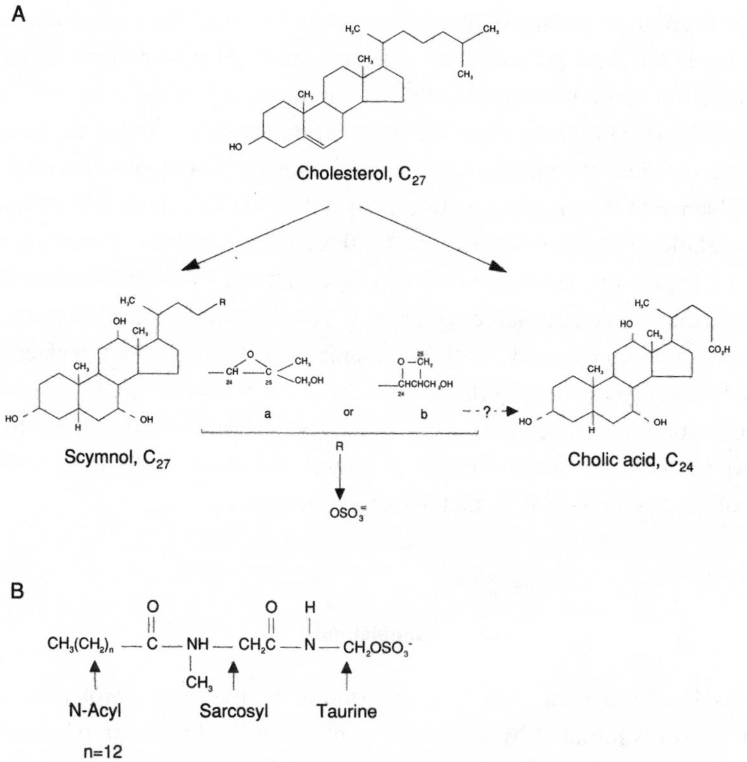

A

Cholesterol, C_{27}

Scymnol, C_{27} a   or   b   Cholic acid, C_{24}

$OSO_3^=$

B

$$CH_3(CH_2)_n - \overset{\overset{O}{\|}}{C} - NH - \underset{\underset{CH_3}{|}}{CH_2C} - \overset{\overset{O}{\|}}{\underset{}{}} N - CH_2OSO_3^-$$

N-Acyl   Sarcosyl   Taurine

n=12

Fig. 2.6 A. Conversion of cholesterol to biliary detergents. During cholic acid formation three terminal carbon atoms of the cholesterol side chain are oxidatively removed ($C_{27} \rightarrow C_{24}$), and three additional OH groups are introduced into the ring system.

The formation of scymnol leaves the C_{27} skeleton of cholesterol intact, but the side chain is oxygenated. In the ring system of scymnol the position of the OH groups is the same as in cholic acid. That cholic acid is formed by way of scymnol is unlikely. Cholic acid and related C_{24} compounds acquire detergent properties by coupling with glycine or taurine to form "bile salts." B. The "bile salt" from cab gastric juice acquires detergent properties by linkage to a —SO_3— residue.

Still another variant of naturally produced emulsifiers or detergents, probably the most primitive, has been encountered in some invertebrates. Daniellson and collaborators (1965), searching for bile acids in crabs and crayfish, noted that the crustacean's gastric juice was foaming vigorously. When purified, the emulsifying material showed no obvious chemical resemblance to the mammalian bile acids but rather to Tide or other modern household detergents (Fig. 2.6B). Its three constituent parts were taurine (see Chapter 12), sarcosine (N-methylglycine) and a ten- or twelve-carbon fatty acid, (fatty acyl-sarcosyltaurine). Thus, the crustacean's detergents share some components with the mammalian bile salts but replace the cholesterol component with a fatty acid, a much simpler lipid. One might argue that for crustaceans the use of a fatty acid rather than a cholesterol-derived molecule constitutes an economy measure, that spares cholesterol, an essential growth factor for invertebrates.

Squalenoids

As we have seen, squalene cyclization to the cholesterol precursor lanosterol is initiated by an oxidative process, the formation of squalene oxide (Fig. 2.1). On purely chemical grounds, squalene might also, and does indeed, cyclize nonoxidatively by protonation rather than by oxygenation (Fig. 2.7). In fact, numerous squalene cyclization products known as pentacyclic triterpenes, which use an anaerobic mechanism throughout, are found in plants and some protozoans. In these instances no flexible side chain remains, as in the case of sterol. One early discovered example is tetrahymanol (Fig. 2.7), isolated from the ciliated protozoan *Tetrahymena pyriformis* (Connor et al., 1971). Cholesterol or any other tetracyclic sterols are absent from this organism and are not needed. *Tetrahymena** is in fact a notable exception to what was once an article of faith, that all eukaryotic cells contain cholesterol (or related tetracyclic sterols). An obvious question is whether squalene-derived tetrahymanol plays a phys-

*Very recently, *Tetrahymena* played a leading role in the discovery of "self-splicing RNA," the unexpected phenomenon that revealed RNA's catalytic (enzymatic) potential (T. Cech and K. Altmann, 1986).

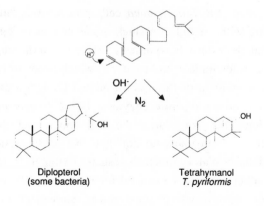

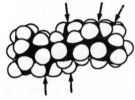

Fig. 2.7 Structure formulas and, below, space-filling models of tetrahymanol and the hopanoid diplopterol. These polycyclic triterpenes are formed anaerobically from squalene directly. Their OH groups are derived from water. The dark arrows point to the α-faces (above) and β-faces (below).

iological role in *Tetrahymena,* akin to that of cholesterol elsewhere in eukaryotes. And indeed this does seem to be the case.

When inserted into artificial or model membranes, tetrahymanol, like cholesterol, raises viscosity and thereby reinforces the membrane. Even more convincingly, *Tetrahymena* cells grown in culture, when supplemented with cholesterol, cease to produce tetrahymanol and contain exogenous cholesterol instead. This replacement occurs without impairment of growth. By all criteria the pentacylic tetrahymanol and the tetracyclic cholesterol are functionally equivalent in the protozoan. Inspection of the three-dimensional structure in Fig. 2.7 reveals—surprisingly—that tetrahymanol shares with cholesterol the feature, stressed above, of displaying two planar regions (α and β faces). This trait explains the evident fitness of tetrahymanol as a membrane-reinforcing insert for the protozoan.

It should be noted that *Tetrahymena* cells produce this "membrane-friendly" structure with the aid of a single squalene-converting enzyme. We have seen that cholesterol biosynthesis, starting with the same hydrocarbon precursor, leads initially to lanosterol, which needs to be streamlined by oxidative removal of three extra methyl groups, a process requiring probably a dozen separate enzymes. But *Tetrahymena's* parsimony probably limits the protozoan to the single function of serving as a membrane stabilizer. Protozoans do not, and one might argue cannot, produce the various regulatory molecules, such as vitamin D, the steroid hormones, and other molecules essential for metazoan function. Hence their classification as protozoans (earliest) can be rationalized on biochemical as well as on structural or morphological grounds.

Bacterial Squalenoids

Some twenty years ago, chemists interested in paleontology began a search for organic molecules in fossils that might have survived exposure to the rigorous conditions of a primitive biosphere. A new chapter on squalenoid chemistry began. Pentacyclic triterpenes named steranes, structurally related to tetrahymanol, were found to be common in ancient sediments and shales, such as the gunflint shards dated to an era one billion to two billion years ago. One can reasonably assume that they represent molecular fossils, the molecular record of ancient organisms that may have produced sterols or, more likely, pentacyclic triterpenes.* Did such molecules occur in primitive forms of life, especially the procaryotic bacterial cells?

Squalene, the ultimate precursor of both sterols and pentacyclic triterpenes, had been isolated earlier from numerous bacteria, yet in these organisms the hydrocarbon remained a molecule of unknown function. One existing clue was the identification of so-called hopanoids, isolated first from ferns, mosses, and trees and abundant in tropical Malaysian rain forests. Hopanoids are named after their discoverer, the British plant

---

*Among the few organic survivors in fossils, rocks, or the fossil fuels coal and petroleum, steranes, are stable at temperatures as high as 400–500°C.

physiologist John Hope. A typical hopanoid structure, diplopterol, is shown in Fig. 2.7. By 1987 some three dozen bacterial species or strains had been shown to contain hopanoids, while in an equally sizable number of others—including *E. coli* and *pseudomonads*—hopanoids are absent (Ourisson et al., 1979). Bacterial taxonomy provides no obvious clues, leaving the role of the bacterial hopanoids uncertain. They have been referred to as cholesterol surrogates or substitutes, but this terminology has only modest experimental support. The growth of some cholesterol-requiring mycoplasmas (bacteria lacking a cell wall) is supported, but only marginally, by the hopanoid diplopterol.

Whether or not hopanoids and related pentacylic triterpenes such as tetrahymanol are functional antecedents of sterols, their formation represents an early and marginally successful device for transforming the flexible hydrocarbon chain of squalene into a rigidly locked structure. Of signal importance for the evolution of squalene-derived molecules, hopanoid formation does not require atmospheric oxygen. Notably also, hopanoids, unlike lanosterol or cholesterol, fail to undergo any subsequent oxidative metabolic transformations of the ring system to hormones and the like. In the context of the modern evolution of the sterol molecule, they are dead-end structures.

A bacterial organism that may have played a landmark role in squalenoid evolution is the previously mentioned obligate methanotroph *Methylococcus capsulatus,* the bacterium that uses and oxidizes methane (marsh gas) as the sole carbon source. Two pathways, both starting from squalene, can be seen at work in this so-far-unique procaryote. *Methylococcus* extracts, when supplied with squalene, convert the hydrocarbon directly to the pentacyclic hopanoid diplopterol by the single-step anaerobic route (Fig. 2.7). When the precursor substrate is squalene oxide, the same bacterial extracts cyclize it to lanosterol and proceed along the cholesterol pathway, demethylation to the intermediates 4,4-dimethyl and 4-methyl sterol, but stop at this stage. Only in *Methylococcus* do the two branches, one to hopanoids and the other to sterols, coexist. Whether the initial events of the two branches are catalyzed by the same enzyme is not clear and awaits purification of the respective enzymes; that research is now under way.

Equally intriguing is the issue of whether the existence of dual pathways that start with the same precursor confers any evolutionary benefits on *Methylococcus*. It probably does not, because hopanoids and sterols appear to be functionally identical, at least qualitatively. I suggest that in *Methylococcus* the anaerobic route of squalene to hopanoids may be vestigial or perhaps redundant. What makes *Methylococcus* so remarkable is that this evolutionary advance, if that indeed is what it is, can be studied in the same organism. Morphological characters strengthen the view that *Methylococcus* may indeed signify an evolutionary turning point. Intracytoplasmic membranes, the kind characteristic for eukaryotes but absent in other prokaryotes, are prominent in *Methylococcus* cells.

Methylococcus capsulatus, the microbe that may have invented several of the steps in sterol synthesis, provides one further opportunity for inquiring into the chronology of the process. Oxygen is an essential reactant for four of the participating steps, for the formation of squalene oxide and for the three sequential demethylations of lanosterol. *Methylococcus* accumulates only two of the demethylated intermediate precursors of cholesterol. Relevant here is the general belief that under primitive earth conditions the terrestrial atmosphere was essentially devoid of oxygen. Early on, the environment was reducing, containing principally hydrogen, nitrogen, methane, and probably CO and CO_2. Not until the advent of the blue-green algae (cyanobacteria) two billion to three billion years ago did the atmosphere become "oxygenic." These primitive cells are credited with the invention of photosynthesis, the light-dependent photolysis of water to $O_2 + H_2R$. Furthermore, it is reasonable to assume that photosynthetic oxygen accumulated in the atmosphere gradually, starting in an ecological niche, spreading, and reaching the current level of 20 percent oxygen at a point of time we cannot yet date or relate to any chemical innovations.

H. P. Klein and his associates have carried out some revealing experiments bearing on the gradual enrichment of atmospheric oxygen (see Jahnke and Klein, 1986). They measured the minimal and optimal oxygen tensions required for the sequential lanosterol demethylation reactions with intact *Methylococcus* cells. The same laboratory has also quantified the oxygen needs of the squalene epoxidase prepared from yeast extracts. The following remarkable results were obtained (shown also in the right-hand column of Fig. 2.3):

1. The yeast enzyme catalyzing the formation of squalene oxide, the earliest oxygen-requiring reaction, operates with an efficiency of 70 percent at 0.03 percent atmospheric oxygen
2. The first lanosterol demethylase of *Methylococcus* producing 4,4'-dimethyl cholesterol requires 0.2–0.3 percent O_2 for optimal activity.
3. For the second demethylase converting 4,4',-dimethyl to 4-monomethyl cholesterol, the oxygen requirements are still higher, about 0.5 percent.

Methylococcus fails to perform the third, final demethylation step, so we do not know what the oxygen requirement might be. Some less direct experiments suggest that it may be even higher, about 1.0 percent.

At any rate, the above results allow the conclusion that the stepwise progression of the sterol pathway from squalene via lanosterol demethylation was driven by the gradually rising oxygen levels in the terrestrial atmosphere. It is equally possible that oxygen levels played a critical role in the induction of the respective enzyme. Most gratifying is the concurrence of the results obtained by Klein's laboratory with the established sequence of steps in the pathway from squalene to cholesterol. Also, the documentation that each sequential demethylation results in functional improvement of the sterol molecule gains strength as an evolutionary argument. Finally, we may ask whether the rapidly accelerating rate of organismic evolution during the transition from prokaryotic to eukaryotic cells (known as the precambrian burst) parallels or perhaps is a consequence of sudden oxygen enrichment of the atmosphere. Many of the typically eukaryotic molecules (hormones, for example) contain oxygen derived from the atmosphere (see Chapter 3). Unfortunately, the oxygen tensions required for most of the respective oxygenases are unknown. Here is a vast subject that the biochemical community might usefully address and thereby contribute to an unsolved evolutionary problem.

I close by reproducing a graph that suggests an exponential rise in the rate of organismic evolution (Fig. 2.8; Schopf, 1970). Below in the same figure are estimates of the ascending concentration of atmospheric oxygen during the same period (Han and Runnegar, 1992). Early on, organismic evolution was slow. The rate began to accelerate dramatically one billion to two billion years ago. Is this rise correlated with the appearance and accelerating accumulation of oxygen in the atmosphere?

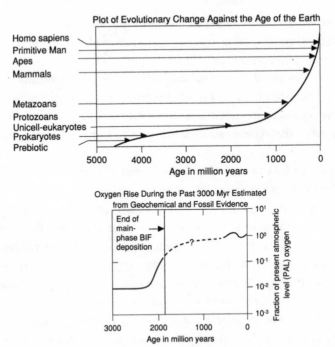

Fig. 2.8 Plot of organismic evolution against the age of the earth (Schopf, 1970) and, below, of hypothetical oxygen tensions in the atmosphere (Han and Runnegar, 1992). BIF stands for banded iron formation.

What started out decades ago as a strictly biochemical project, the mechanism of cholesterol biosynthesis, has broadened to a wide range of problems, including the beginnings of life on this planet.

Bibliography

Comprehensive Texts

1. L. Fieser and M. Fieser (1959), *Steroids*. New York, Reinhold Publishing Corporation.

2. W. Nes and M. McKean (1977), *Biochemistry of Steroids*. Baltimore, Maryland, University Park Press.

Oxygen and Sterol Biosynthesis

3. T. T. Chen and K. Bloch (1956), Mechanism of squalene cyclization, *J. Am. Chem. Soc.* 78, 1517.

4. A. Eschenmoser, L. Ruzicka, O. Jeger, and D. Arigoni (1955), Eine sterochemische Interpretation der biogenetischen Isoprenregel bei den Terpenen, *Helv. Chim. Acta* 38, 1890.

5. E. J. Corey et al. (1966), The formation of squalene 2,3 oxide, *J. Am. Chem. Soc.* 88, 4750.

6. E. Van Tamelen at al. (1966), Squalene 2,3 oxide conversion to cholesterol, *J. Am. Chem. Soc.* 88, 4752.

Demethylation of Lanosterol to Cholesterol

7. K. Bloch (1983), Sterol structure and membrane function, *CRC Crit. Rev. Biochem.* 14, 47.

8. C. Dahl, J. Dahl, and K. Bloch (1988), Properties of mycoplasma membranes cultured on cholesterol, *Biochemistry* 19, 1467.

9. C. Dahl and J. Dahl (1988), *Cholesterol and cell function in biology of cholesterol,* ed P. L. Yeagle, p. 147. Boca Raton, Florida, CRC Press.

Cholesterol-Derived Detergents

10. G. Haslewood (1952), Biochemical studies of bile salts, *Biochem. J.* 52, 583.

11. A. van der Oord, H. F. Daniellson, and R. Ryhage (1965), Structure of the emulsifier in crab gastric juice, *J. Biol. Chem.* 240, 2242.

Sterols of Insects and Invertebrates

12. J. Clark (1958), Metabolism and function of sterols in insects, Doctoral dissertions, Harvard University.

13. C. Djerassi et al. (1979), Recent progress in the marine sterol field, *Pure Appl. Chem.* 51, 1815.

Steroids and Hopanoids in Bacteria

14. C. W. Bird et al. (1971), Steroids and squalene in *Methylococcus capsulatus, Nature* 230, 473.

15. G. Ourisson, P. Albrecht, and M. Rohmer (1979), The hopanoids: Palaeochemistry of a group of natural products, *Pure Appl. Chem.* 51, 709.

16. R. Connor et al. (1971), Tetrahymanol in *Tetrahymena* membranes, *Biochem. Biophys. Res. Comm.* 44, 995.

17. C. Egliniou and G. B. Curry, eds. (1991), *Molecules through fossil time, fossil molecules and systematics*. London, Royal Society.

Oxygen and the Precambrian Explosion

18. L. D. Jahnke and H. P. Klein (1986), Methylsterol composition of *Methylococcus capsulatus, J. Bact.* 167, 238.

19. J. W. Schopf (1970), Precambrian microorganisms and evolutionary events prior to the evolution of higher plants, *Biol. Rev.* 45, 319.

20. T. M. Han and B. Runnegar (1992), Microscopic algae from the 2.1-billion-year-old negaunee iron formation, Michigan, *Science* 257, 232.

3

Oxygen and Evolution

_____

After the advent of atmospheric oxygen during evolution,

a vast array of biochemical and morphological

innovations occurred in the organic world. An example is

the oxygen-dependent formation of diverse metabolic

signals, all eukaryotic.

The advent of oxygen in the terrestrial atmosphere was the second major event in the evolution of organisms, subsequent to the origin of life. It is commonly believed that photosynthetic blue-green algae several billion years ago were the first cells to produce atmospheric oxygen by photolysis of water; but some earlier inorganic, geochemical event cannot be ruled out, at least as a source of minor amounts of oxygen. How rapidly the concentration of atmospheric oxygen reached the current level of 20 percent is also unknown. The issue is not trivial because some aerobic reactions occur at much lower oxygen tensions than others (Chapter 2).

Most discussions on this subject are limited to the profound benefits gained by aerobic cells relative to those that exist in the absence of atmospheric oxygen. Aerobic cells respire, (that is, they can convert all of the carbon atoms of sugars or fatty acids to carbon dioxide and water), producing up to fifteen times as much chemical energy in the form of

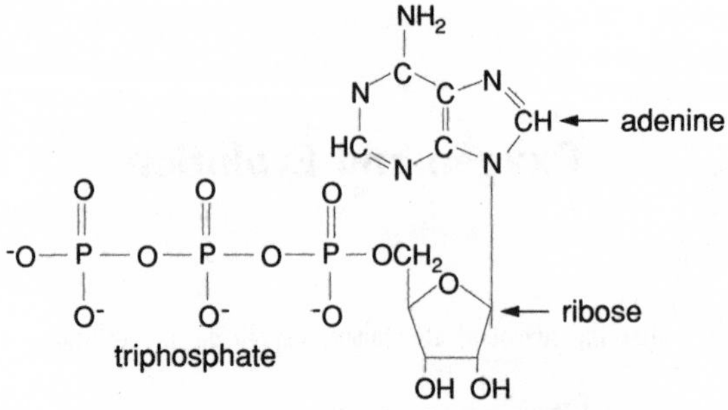

Fig. 3.1 Adenosine triphosphate (ATP).

adenosine triphosphate (ATP) as anaerobic cells (Fig. 3.1). In the absence of oxygen, organisms such as yeast and anaerobic bacteria convert sugar to lactic acid and ethanol, extracting only a fraction of the energy available from total combustion of a carbon source. The bulk of the oxygen, probably more than 90 percent consumed by aerobic cells serves as the terminal electron acceptor in respiration. Equally essential for aerobes, oxygen serves as an obligatory reactant for a variety of biosynthetic energy-consuming reactions.

Although Lavoisier discovered respiration two hundred years ago, the second category of oxygen-requiring biosynthetic reactions was recognized much later. Postulated early in the twentieth century, they were not demonstrated experimentally until the 1950s by O. Hayaishi and H. Mason (1955). Isotopic oxygen, $^{18}O_2$, was essential for recognizing processes now know as oxygenase reactions (Fig. 3.2). One early example is the origin of the OH group in cholesterol, discussed in Chapter 2.

In a different context (Chapter 11) I mention Pasteur's famous experiments on yeast fermentation, which led to what he called "la vie sans air." It does not lessen or detract from the importance of Pasteur's dictum that

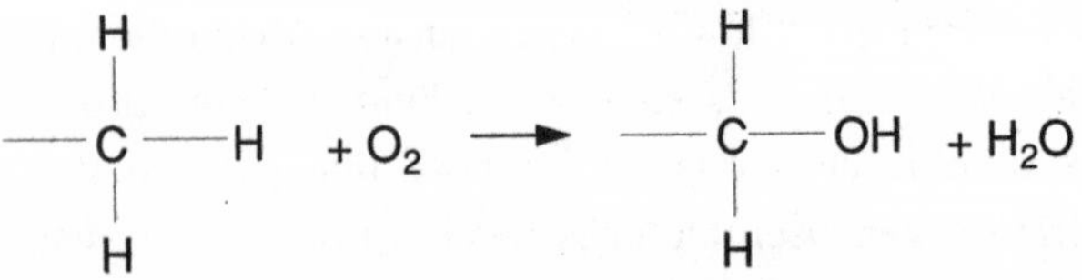

Fig. 3.2 Oxygenase reactions catalyzing the entry of molecular oxygen into organic compounds.

his experimental system—without his knowledge—contained traces of oxygen. We now know that yeast belongs to the category of microaerophilic organisms that require minimal oxygen tensions, sufficient for oxygenase reactions but not for respiration. Strictly anaerobic organisms do exist, but yeast is not among them. True anaerobes are found only in the bacterial kingdom—the prokaryotes, presumed to be the most ancient forms of life, and primitive both structurally and biochemically.

Given the fact that life exists in the complete absence of oxygen, it follows that the minimal biochemical processes essential for life were necessarily anaerobic: the multitude of synthetic reactions producing amino acids and proteins, the purine and pyrimidine bases for nucleic acid synthesis, the lipid constituents of membranes, and many of the coenzymes. Remarkably, and a fact that is rarely emphasized, the overall mechanism and the basic chemistry of these minimal life processes remained unchanged when oxygen appeared in the atmosphere. This retention of primitive pathways throughout the evolution of organisms argues of course powerfully for the unity or universality concept of life processes.

In 1959, soon after oxygenases were discovered, J. R. Nursall, in a paper entitled "Oxygen as a Prerequisite to the Origin of Metazoa," pointed out that "less attention has been focused on the next steps in organic evolution, namely those attending the diversification of primitive living forms into the kingdoms of organisms recognized today," and "I wish to argue that the rapid diversification was brought about by one ecological change, namely by the addition to the environment of a surplus of pure oxygen."

The key word here is "diversification," today known as biodiversity. A few years later, Howard Goldfine and I (1963) summarized the rapidly growing evidence for a chemically based diversity of the forms of life, resulting from secondary adaptations involving either loss or gain of chemical abilities superimposed on the fundamental or universal ground plan. The appearance of atmospheric oxygen was the driving force behind evolutionary innovations.

I shall make several points in this chapter. First, many of the products of oxygen-requiring reactions are unique to the aerobic lifestyle. They are absent in anaerobes and therefore represent metabolic and functional specialization. Second, universal cell constituents are not invariably formed by single pathways—as the principle of biochemical unity would require. Certain molecules are synthesized anaerobically in some organisms and aerobically in others, by entirely different routes.

There is a general belief, necessarily no more than a belief, that life arose in an anaerobic environment. The early atmosphere of the earth was reducing. It could hardly have been otherwise. Given that the chemicals necessary for building the constituents of primitive cells were formed prebiotically, the presence of oxygen would have precluded their accumulation. Exposure of organic compounds to oxygen would have destroyed the organic substances dissolved in the rich organic soup postulated and popularized in the 1920s by A. I. Oparin and J. B. S. Haldane.

Synthesis of Olefinic Acids

My interest in this subject dates from the observation that, in yeast, oxygen is needed for transforming stearic acid to oleic acid, and from a subsequent inquiry into corresponding mechanisms of anaerobic bacteria.

The basic features of the two pathways, aerobic and anaerobic, are as follows (Fig. 3.3A):

In yeast, animal tissues, plants, and some aerobic bacteria, olefinic acids, such as oleic acid are formed by hydrogen abstraction from two adjacent carbon atoms of the corresponding saturated acid. Oxygen and a reducing agent (NADPH) and cytochrome b_5 (see below) are required for this reaction. Thioesters of coenzyme A (COSR) are the "activated" substrates. Precursor and desaturation products have the same length (eighteen carbon atoms).

In the anaerobic pathway (Fig. 3.3B) the precursors for olefin formation are acids (thioesters) of medium chain length, for example C_{10}. As we will see in greater detail when discussing the chain lengthening process (Chapter 7), several successive C_2 additions yield β-keto acids. At each stage these intermediates are subsequently reduced to β-hydroxy acids. The oxygen of hydroxy acids ultimately comes from water, not air. Removal of the elements of water then transforms the hydroxy derivative into an olefin. Subsequently, several C_2 units are added, with retention of the double bond to form a long-chain analog of oleic acid. The product retains

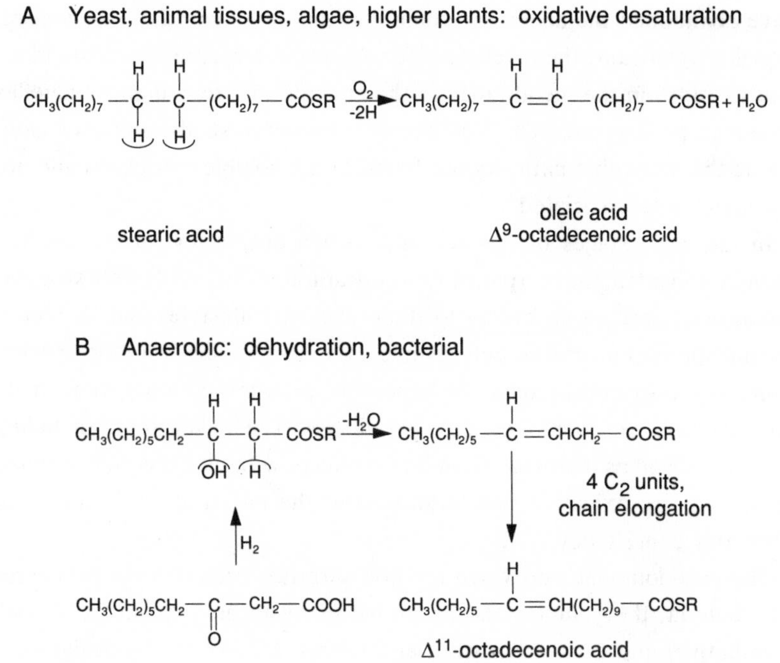

A   Yeast, animal tissues, algae, higher plants: oxidative desaturation

$$CH_3(CH_2)_7 - \overset{\overset{H}{|}}{\underset{\underset{H}{|}}{C}} - \overset{\overset{H}{|}}{\underset{\underset{H}{|}}{C}} - (CH_2)_7 - COSR \xrightarrow[-2H]{O_2} CH_3(CH_2)_7 - \overset{\overset{H}{|}}{C} = \overset{\overset{H}{|}}{C} - (CH_2)_7 - COSR + H_2O$$

stearic acid

oleic acid
Δ^9-octadecenoic acid

B   Anaerobic: dehydration, bacterial

$$CH_3(CH_2)_5CH_2 - \overset{\overset{H}{|}}{\underset{\underset{OH}{|}}{C}} - \overset{\overset{H}{|}}{\underset{\underset{H}{|}}{C}} - COSR \xrightarrow{-H_2O} CH_3(CH_2)_5 - \overset{\overset{H}{|}}{C} = CHCH_2 - COSR$$

4 C_2 units,
chain elongation

$\uparrow H_2$

$$CH_3(CH_2)_5CH_2 - \overset{}{\underset{\underset{O}{\|}}{C}} - CH_2 - COOH \qquad CH_3(CH_2)_5 - \overset{\overset{H}{|}}{C} = CH(CH_2)_9 - COSR$$

Δ^{11}-octadecenoic acid

Fig. 3.3 Comparison of mechanisms for converting saturated to unsaturated
(olefinic) fatty acids.

the double bond located two carbon atoms farther from the carboxyl end
than in oleic acid. No coenzymes are needed. To underline the essential
differences of the two pathways, oxygen-assisted removal of hydrogen is
the olefin-generating event in the aerobic pathway (desaturation), and
removal of water (dehydration) achieves the same end in anaerobic cells.

As for physical properties, the products of the two pathways, the Δ^9
and Δ^{11} acids, appear to be identical. Both melt at 15° C. Metabolically,
however, they are not interchangeable.

One generalization regarding the occurrence of the two pathways can
be made with reasonable certainty: in no instance thus far do the aerobic
and anaerobic processes exist side by side.* Aerobic organisms appear to

*Already this statement is no longer true. A recent paper (Morita et al.,
1992) reports that "the anaerobic pathway and aerobic desaturation both occur
in the synthesis of unsaturated fatty acids during aerobic growth of Vibrio

have abandoned the anaerobic chain elongation-dehydration mechanism. Equally significant, the machinery for oleic acid formation in most eukaryotes resides in the membranes of the endoplasmic reticulum, organelles absent in prokaryotic cells.* In bacteria, by contrast, the enzymes catalyzing the anaerobic pathway are found in the soluble cytoplasm and are not membrane associated.

In the early stages of this research, when only a few organisms had been investigated, the two pathways to unsaturated fatty acids were straightforwardly classified according to the organisms' lifestyles and structural organizations: eukaryotic cells employ the aerobic desaturation mechanism, while in prokaryotes the anaerobic dehydration-elongation mode prevails. Matters were not quite so simple. The eukaryotes, including protists as well as metazoa, do in fact produce, without exception, olefinic fatty acids by oxidative desaturation. But the bacterial world does not show this consistency.

The pseudomonads provided the first surprise. Although obligate aerobic bacteria, they employ the anaerobic pathway first shown for *E. coli* (facultative) and *Clostridia* (strict anaerobes). Moreover, the designation of the anaerobic pathway as bacterial and of the oxidative pathway as typically eukaryotic became untenable when oxidative desaturation of stearate to oleate was shown to be the pathway in some mycobacteria. These bacteria, classified as actinomycetes, and including for example tubercle bacilli, typically display a mycelial (mushroom-like) pattern of growth, in contrast to *E. coli* or the strictly anaerobic *Clostridia*. Moreover, and probably more relevant, the cofactor requirement for the mycobacterial desaturase is similar to if not identical with the corresponding enzymes of yeast and animal tissues. Equally significant, the mycobacterial desaturase is associated with membranous elements, not water soluble. Clearly, aerobic desaturation was not an eukaryotic invention.

The question remains whether the development of the aerobic path to olefinic acids was a unique event in evolution or whether it arose inde-

----

strain ABE-1." The lesson is that generalizations in biology should always be qualified—if not avoided altogether.

*Exceptions to the rule are the water-soluble and oxygen-dependent desaturases of higher plants (spinach chloroplasts and the phytoflagellate *Euglena*).

pendently more than once. Comparing the amino acid sequences of desaturases, (for instance, from mycobacteria) with those of yeast and animal tissues, and determining whether or not homologies exist among the enzymes from different sources, might answer this question.

There appears to be a distinct division between the fate of monounsaturated fatty acids synthesized by the prokaryotic world and those synthesized by the eukaryotic world. Prokaryotes produce unsaturated fatty acids containing no more than a single double bond. In animals and in plants, the desaturation process continues with insertion of additional double bonds along the fatty acid chain to produce in turn di-, tri-, tetra-, penta-, and hexaenoic acids (Chapter 12). Whether these processes require new cofactors or coenzymes is not known. At any rate, they appear to be true innovations that are of importance not only in themselves but as precursors for hormonal messengers common to eukaryotic cells. The prostaglandins and leukotrienes formed from the tetraenoic arachidonic acid by oxygenase reactions are the most notable examples, certainly landmark occurrences of the protozoan and metazoan world.

All cells, whether prokaryotes or eukaryotes, require olefinic acids as membrane constituents for physical reasons. Fluidity is the important parameter, and in most instances this fluidity is achieved by insertion of one or more double bonds into saturated fatty acid chains. Having made this dogmatic statement, I must immediately qualify it. There is yet another device used by Nature to achieve fluidity. The archaebacteria, a phylogenetic line of prokaryotes believed to be distinct from the eubacteria and presumed to be more ancient (C. Woese), possess branched chain

lipids
$$\begin{matrix} & H \\ C-&\!\!\!\!C\!\!\!\!&-C \\ & | \\ & CH_3 \end{matrix}$$
in lieu of olefinic chains
$$\begin{matrix} & H\ \ H \\ -C-&\!\!\!\!C\!\!=\!\!C\!\!\!\!&-C- \end{matrix}$$. Nature

has certainly tinkered with different solutions for what superficially seem to be identical functions.

Oxygen and Amino Acid Biosynthesis

The formation of all twenty amino acids, the building stones for protein biosynthesis and one of the fundamental and universal processes, proceeds in the the total absence of oxygen. The aerobic mode of life has not

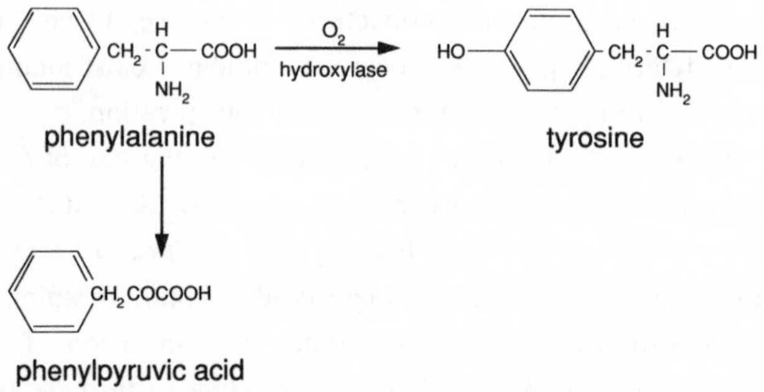

phenylalanine   tyrosine

phenylpyruvic acid

Fig. 3.4 Conversion of phenylalanine to tyrosine and phenylpyruvic acid.

changed or modified that process or added to the repertoire, with one possible but probably irrelevant exception. In animals the essential amino acid phenylalanine (from plant sources) can be oxygenated or hydroxylated to tyrosine (Fig. 3.4). This reaction requires a coenzyme known as tetrahydrobiopterin, which is apparently absent in bacteria.

Since all normal diets contain tyrosine, this amino acid, unlike phenylalanine, is not essential. On the contrary, when the enzyme phenylalanine hydroxylase is absent, an excessive accumulation of phenylalanine ensues, directing the amino acid into alternate metabolic routes. One of these is a loss of the amino nitrogen to form the keto compound phenylpyruvate (Fig. 3.4). This derivative passes into the urine, a condition known as phenylketonuria (PKU). An inborn error of metabolism, PKU is associated with severe mental retardation of unknown origin. A cure for this genetic disorder does not yet exist, but rigorous restriction of dietary intake of phenylalanine will alleviate the symptoms. Untreated infants have an IQ in the low fifties, which returns to normal on phenylalanine-restricted diets.

A somewhat related but benign human disorder ensues when the further oxidative metabolism of tyrosine is blocked. The disorder, known as alkaptonuria, causes the urinary excretion of homogentisic acid, a catabolite that polymerizes spontaneously to products that blacken the urine. The discovery of alkaptonuria by Sir Archibald Garrod in 1902 is one of the milestones in genetics. For the first time a gene was implicated in a metabolic event. In Garrod's words, "Splitting the benzene ring [of tyrosine] in normal metabolism is caused by a special enzyme that in congenital alkaptonuria is wanting." Thus the direct relation between genes and enzymes was recognized, and the concept and proof of "inborn errors in metabolism" was established.

In a sense, these two examples show the potentially adverse effects of oxygen in situations when a step in the normal oxidative metabolism of an amino acid is blocked. The literature fails to mention whether alkaptonuria and phenylketonuria occur in any animal species, including primates, other than man. As for prokaryotic organisms capable of oxidizing phenylalanine to tyrosine, the obligate aerobic pseudomonads can carry out this reaction. Of late, some pseudomonad species have assumed special importance by virtue of promising relief from environmental disasters. These bacteria are now in use, with some success, to convert spilled oil into water-soluble products. They not only oxidize phenylalanine to tyrosine but also attack a variety of "aromatics," including derivatives of benzene and toluene found in petroleum.*

Albinism

We have said that the amino acid tyrosine is subject to a variety of oxidative modifications, leading to hormones (adrenaline, thyroxine) and in some genetic disorders (alkaptonuria, phenylketonuria) to deleterious products. Below is a listing of other oxygen-dependent metabolic innovations.

A. *Lipids*
Stearic acid → oleic acid
Oleic acid → multi polyunsaturated acids; linoleic, linolenic, and arachidonic acids

---

*Benzene and toluene are toxic to humans, and possibly carcinogenic. In the past, liter quantities of these volatile solvents were widely used in routine chemical operations. What currently lessens this potential health hazard is the substantially reduced scale, probably hundredfold or more, of chemical operations. Vastly improved, more sensitive analytical procedures developed in the last several decades make this possible.

Arachidonic acid → prostaglandins → prostacyclin
thromboxan, Leukotrienes,

Sterol Biosynthesis:
Squalene → squalene epoxide
Demethylation of lanosterol
Cholesterol → bile acids

Hormones from Cholesterol:
Androgens, estrogens, corticosterols
Ecdysons (insect developmental hormones)
Cholesterol → vitamin D → 1,25-dihydroxycholecalciferol

β-carotene → vitamin A → retinal → rhodopsin

B. *Amino Acids*
Proline → hydroxyproline
$\qquad$ | → collagens
Lysine → hydroxylysine

Tyrosine → dopa, dopamine
Tyrosine → norepinephrine, epinephrine (adrenaline)
Tyrosine → melanins, hair pigments
Tyrosine → thyroxin, thyroid hormone

Tryptophan → nicotinamide
Tryptophan → serotonin
Tryptophan → melatonin (pineal gland)
Tryptophan → indoleacetic acid (plant growth hormone)

Porphyrinogens → porphyrins (iron-porphyrins, heme cytochromes,
Mg-porphyrins, chlorophylls

The most common genetic defect of tyrosine metabolism is albinism, a
generic designation comprising diverse clinical syndromes. The cause is
hypomelanonis, the inability of pigment cells (melanocytes) to convert
tyrosine to melanin. Melanin (from *mela,* black) is perhaps a misnomer.
As commonly used, it refers to the pigments of the skin, eye, and hair of
all shades from black to brown, red to the lightest yellow.

The melanin-forming enzyme tyrosine oxidase is multifunctionai, catalyzing at least six consecutive oxidative conversions of tyrosine, followed by spontaneous polymerization to a high-molecular, still uncharacterized product.

Two distinctive melanins result from tyrosinase action, pheomelanins (*pheo,* for red) and eumelanins, standing for light brown. Pliny described albinism as early as the first century A.D., and some passages in the Dead Sea scrolls indicate that Noah was an albino. To date, fourteen different types of albinism have been described in humans, four affecting the color of the eye (ocular albinism, OA), ten affecting both ocular and skin pigments (ocular-cutaneous, OCA). There are yellow mutant albinos who have yellow-red hair; their hair bulbs do not form the black eumelanin. The hair of platinum albinos is a metallic-cream platinum color; still others, such as minimum-pigment albinos, have hair that is stark white. Along with the various types of albino hair goes blue or gray-to-blue eye color, which in some instances darkens with age.

An early description (1699) of albinos encountered among the Cuna Indians inhabiting an island near Panama reads as follows: "There is one complexion so singular . . . that I never saw elsewhere in any part of the world . . . They are a milky white, much like the color of a white horse. Seeing them in a moonlit night, we used to call them moon ey'd."

Sir Archibald Garrod, the discoverer of the genetic basis of metabolic disorders, in 1907 was led to speculate that albinism was an inborn error of metabolism—like alkaptonuria, also due to impaired tyrosine metabolism. In humans, the principal symptom of albinism is photophobia due to toxicity. Because of weak or absent tyrosinase activity and hence impaired melanin formation, the skin of albinos will not tan and form the normal protective pigmentation.

In the world's population the prevalence of albinism is 1 in 20,000 varying in frequency from as few as 1 in 180,000 for some caucasians to as many as 1 in 100 in some Mexican tribes. There is no cure for this congenital deficiency. In principle, genetic engineering could develop a therapy by replacing one or more of the tyrosinase genes. This technique, although no longer a dream, lies far in the future and is of low priority because albinism is not a life-threatening disease, except for an increased frequency of skin cancer.

Nicotinic Acid

As described in Chapter 10, one of the essential functions of the amino acid tryptophan is to supply nicotinic acid in animals and plants for synthesis of the coenzymes NAD and NADP. This synthesis is a complex process involving several oxygenase reactions. Necessarily, anaerobic bacteria and perhaps prokaryotes generally lack the requisite machinery for producing the pyridine nucleotides NAD and NADP by oxidative breakdown of tryptophan. They rely on an alternative mechanism, to be discussed in Chapter 8. It is mentioned here as an additional example of replacement of an anaerobic pathway by an oxygen-driven mechanism—with the same end result. Again, there is no evidence that the two mechanisms coexist in a given organism.

Oxygen and Collagen Synthesis

Collagens (from the Greek *colla,* glue) comprise a family of insoluble fibrous proteins essential for reinforcing (that is, conferring high tensile strength on) vertebrate connective tissues. They are structural elements found in bone, skin, blood vessels, and teeth. Structurally, collagens differ in several respects from other proteins such as enzymes. Only the role of oxygen will be considered here. When collagen is hydrolyzed, two of its fifteen constituent amino acids, proline and lysine, are found to contain extra hydroxyl groups not seen in any nonstructural or soluble proteins (Fig. 3.5).

Interestingly, molecular oxygen, the source of these hydroxyl groups, does not enter at the free amino acid stage before collagen synthesis begins, but the OH groups first appear in certain polypeptides known as procollagens, either during or after protein synthesis. Such events are known as posttranslational modifications, a widespread device for altering the properties of proteins after polypeptide chains have been assembled.

Another oxidative transformation of collagen-linked lysine occurs at the terminal or ϵ-NH$_2$ groups of proximal lysine residues (Fig. 3.6). Two such R-CHO groups undergo a coupling reaction known as aldol condensation. Cross-linking—formation of a network, and hence mechanical strength-

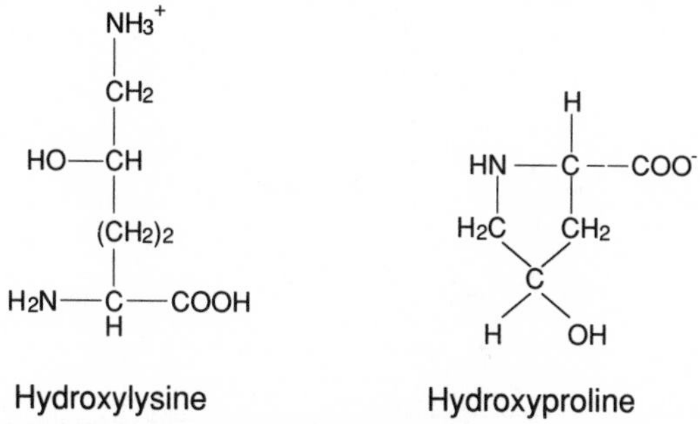

Hydroxylysine   Hydroxyproline

Fig. 3.5 Hydroxylated amino acids contained in
collagen and other fibrous proteins.

ening—is thereby achieved. Yet the more such cross-linking occurs, and
it does increase with advancing age in humans, the more bone collagen
becomes fragile, contributing to the high risk of bone fractures in the
elderly.

One and perhaps the only well-established role of ascorbic acid, vitamin
C, is to serve as cofactor for proline and lysine hydroxylation during
collagen synthesis. Collagen synthesized in the absence of the vitamin
(because it is inadequately hydroxylated) is unable to form the cross-links
that confer mechanical strength on fibrous proteins. Scurvy, the manifes-
tation of extreme ascorbic acid deficiency, was experienced by early sea-
farers before the scurvy-preventing properties of citrus fruits or fresh
vegetables were recognized in the eighteenth century. If, as all of the
evidence suggests, collagen—the most abundant protein in mammalian
species—is a typical product of evolutionary progress from unicellular to
multicellular metazoans, and if ascorbic acid is essential only in collagen
synthesis, then there is no need for this vitamin in unicellular forms of
life. Indeed, procaryotes and unicellular eukaryotes such as yeast do not
contain, nor do they need, ascorbic acid. Obviously, citrus fruits and

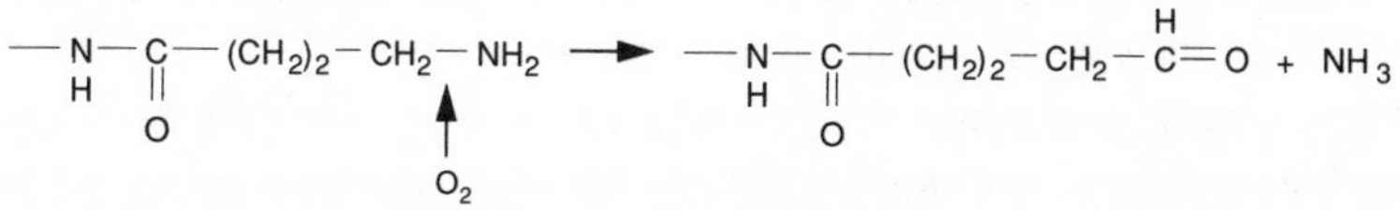

Fig. 3.6 Oxidative transformation of lysine prior to cross-linking in collagen.

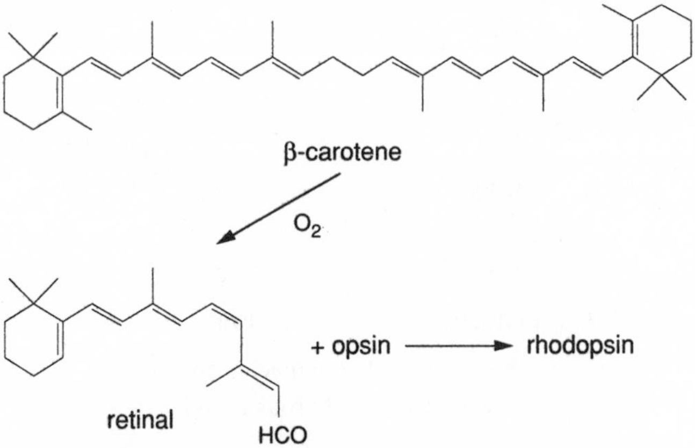

Fig. 3.7 Conversion of the plant pigment β-carotene to rhodopsin, the visual purple of the retina.

vegetables, rich sources of ascorbic acid, are important for some functions—whatever they are—in higher plants.*

Vitamin A

The roles of vitamin A, retinal, and rhodopsin will be dealt with in Chapter 12. Here, in the context of oxygen and evolution, a rather unique case of retinal function deserves mention.

It is clear that development of the visual apparatus followed or coincided with the appearance of retinal and rhodopsin in multicellular eukaryotes. How early or how late did these events occur?

By some mechanism, still unknown** except that it requires oxygen, intestinal cells cleave β-carotene in the central portion to form retinal (Fig. 3.7). In turn, retinal combines with the protein opsin to form rhodopsin,

*While there is some evidence for an ascorbic acid requirement of parasitic protozoans, the role of the vitamin in unknown.

**Potentially, β-carotene (C_{30}) could give rise to two molecules of retinal (C_{15}) but, curiously, the stoichiometry of the cleavage reaction has not been determined. The mechanism of the oxidative cleavage is also unknown.

the light-sensitive molecule in the rods and cones of the retina. Vitamin A deficiency, which results in the failure to form rhodopsin, leads to night blindness and eventually to degenerative changes in the retina.

Presumably unicellular eukaryotes, such as yeast, do not need photoreceptors. Light does not affect their growth. Yet the light-absorbing pigment rhodopsin appears to be an invention of some very early prokaryotic organisms. Certain salt-loving bacteria, for example *Halobacterium cutirubrum,* growing in solutions containing up to 20 percent sodium chloride, use the pigment to convert light energy into ATP. Their purple-colored membrane* contains bacteriorhodopsin and functions as a photoreceptor. The process is not vision in the vertebrate or metazoan sense but perhaps a primitive form of harvesting light for energy production, abandoned by descendants of the halobacteria.

Porphyrins

Early in this chapter we noted that most evolutionists, if they mention aerobic versus anaerobic life forms at all, stress the role of oxygen in energy production in the form of ATP.** The invention of respiration enhanced ATP production by more than an order of magnitude. The molecules ultimately responsible for the combustion of organic substances to CO_2 and H_2O are known as porphyrins (from *porphyra,* purple), or when combined with iron, as hemes, the biologically active form (Fig. 3.8).

The beholder will admire the beautiful symmetry of heme, the molecule essential for the aerobic lifestyle.

The key feature of the porphyrins' structural design is to bind Fe^{++} (in

---

*These purple bacteria have a choice. In shallow seawater at high light intensity, their ATP production is light driven, whereas in the dark, at night, or in deep water, the mechanism of ATP generation is the conventional one, aided by a proton pump.

**An authoritative symposium entitled *Evolution from Molecules to Man* (Cambridge, Cambridge University Press, 1983) makes no mention of the additional roles of oxygen in evolution discussed here.

the ferrous, or reduced, form) and to hold it tightly (in chlorophyll, magnesium takes the place of iron). In the red cell, heme combines with the protein globin to form hemoglobin, the universal oxygen carrier for respiring vertebrate tissues.

$$\text{Hemoglobin } Fe^{++} + O_2 \rightleftharpoons \text{Hemoglobin } FeO_2$$
$$\text{oxyhemoglobin}$$

Hemoglobin serves as a vehicle to supply respiring tissues with oxygen; it is not itself involved in oxidative reactions. In the process the iron atom remains reduced (Fe^{++}) without changing its valence. In turn, tissues adequately supplied with oxygen oxidize organic molecules with the aid of a series of cytochromes, heme-containing electron carriers that successively accept electrons and pass them serially to the terminal member of the oxidation chain. This enzyme chain known as cytochrome oxidase (cytochrome a_3) interacts directly with oxygen, converting it to water:

$$\text{substrate} \rightarrow \rightarrow \rightarrow \underset{\text{electron flow}}{\underline{\text{cyt. b} \rightarrow \text{cyt. c} \rightarrow \text{cyt } a_3 + O_2 \rightarrow}} \text{cyt. } a_3^{ox}$$
$$\curvearrowright 2H_2O$$

All the cytochromes share the same basic iron-porphyrin ring system with the heme component of hemoglobin. Cytochrome oxidase, catalyst for the ultimate transfer of electrons from reduced cytochrome c to oxygen, is a molecule of extreme complexity, not yet fully characterized. It is known to contain two heme groups and two copper atoms. Otto Warburg, the first biochemist to carry out extensive investigations on this elusive molecular assembly ("*Atmungsferment,* or respiratory enzyme"), referred to it in 1931 as a substance "that seems less accessible to man than the surface matter of the moon" (H. A. Krebs, personal communication). Perhaps this is still true today.

The foregoing discussion on respiration—grossly oversimplified—intends to acquaint the reader with the porphyrin structure, the invention of a molecule that must have been present at the beginning of aerobic life, or preceded it. We saw previously that the essential or minimal life processes took place in the absence of oxygen and have remained anaerobic throughout evolution. Porphyrins were not yet invented. However, molecules related in some respects to heme and sharing a common precursor occur in anaerobic bacteria. They are the corrins, basic components of vitamin B_{12}, a coenzyme for a variety of enzymatic reactions.

The corrin structure may be regarded as a modified porphyrinogen nucleus lacking one of the bridging carbon atoms (marked by an asterisk in Fig. 3.8). Moreover cobalt rather than iron occupies the central cavity. Corrins are formed from the porphyrinogen precursor in the absence of oxygen, in contrast to the porphyrinogen $\rightarrow$ heme conversion. The number of double bonds ($=$) in the porphyrinogen-corrin conversion is reduced from eight to six, whereas porphyrin or heme formation introduces four additional double bonds, a transformation that requires molecular oxygen.* Functionally also, the corrins and the porphyrins play entirely different roles. The B_{12} coenzymes are involved in a number of intramolecular rearrangements, notably in the anaerobic conversion of ribose derivatives to deoxyribonucleotides, while the heme-derived cytochromes promote respiratory electron transfer, ultimately to molecular oxygen.

Vitamin B_{12}

Cobalamine, or B_{12}, qualifies as a vitamin in the sense that the mammallian organism cannot synthesize it.** Yet dietary supplements of B_{12} are needed only for the rare patient afflicted with pernicious anemia. Intestinal (anaerobic) bacteria provide the human organism with the vitamin, usually in adequate amounts. On average, the dose for normal adults is one nanogram (1×10^{-9} gm), the smallest recommended dose for any vitamin.

The history of vitamin B_{12} is a fascinating one. Two physicians, George

*There are reports claiming a conversion of porphyrinogen to porphyrins under anaerobic conditions, as in *Rhodopseudomonas spheroides*. Were the experimental conditions truly anaerobic however? In my own experience, complete removal of oxygen is very difficult to achieve; unless the removal has been rigidly controlled, such claims may not be valid.

**Occasionally I stray from a chapter's main subject. Such diversions are ordinally relegated to footnotes. The discovery of B_{12} was so extraordinary, though, that it deserves to be part of this chapter's narrative.

Minot and William Murphy of Harvard Medical School, were treating patients suffering from pernicious anemia, until the 1920s an incurable disease. Minot and Murphy were aware of George Whipple's success (at the University of Rochester) in alleviating a variety of anemias, such as accidental loss of large volumes of blood, by diets supplemented with massive amounts of liver.* Pursuing the cue, Minot and Murphy succeeded in relieving the disease by prescribing massive amounts of raw liver for their patients' diets. The treatment was a spectacular success. What was the active anti–pernicious anemia liver factor? I suspect the physicians faced a dilemma. In order to fractionate the liver factor and follow the progress of purification, they would have to divide the patients into two groups, one receiving the beneficial liver fractions and the other, the control group, kept on a liver-free diet. Was it not unethical to withhold the liver treatment from any of the patients?

Some progress was made with the help of E. J. Cohn, a colleague of Minot and Murphy. He prepared relatively potent liver extracts suitable for intravenous injections to replace the unpalatable liver diet. But the chemistry of the liver factor remained totally unknown. Another obstacle was that the disease could not be produced experimentally in animals. A chance discovery some twenty years later solved that predicament. An anaerobic bacterium, *Lactobacillus dorner,* unable to grow on simple media, responded to the same liver fractions that brought the human disease under control (Shorb, 1948). The bacterial growth test proved to be a godsend for assaying the anti–pernicious anemia liver factor, which was promptly (in fact, a few months later) obtained in pure crystalline form by E. Smith in Britain and K. Folkers in the United States. Structural analysis by several outstanding chemists of the time, and most crucially the x-ray crystallography performed by Dorothy Hodgkin in Britain, revealed a porphyrin-like structure with a crucial difference: during the porphyrinogen → corrin conversion, one of the CH groups connecting the

---

*When I first learned of pernicious anemia, I was told the following probably apocryphal story. One of Minot's patients, when questioned, responded that he was very fond of liver pâté and felt very much better after indulging in this delicacy.

pyrrol rings (starred in Figs. 3.8 and 3.9) is removed, linking the two rings directly rather than by way of a "methine" bridge.

What was the mechanism for removing this methine bridge? Nature provided a clue. One of the intermediates isolated in the course of studies on B_{12} biosynthesis in the anaerobic bacterium *Propionic bacterium shermani* contained an extra methyl group that the final B_{12} vitamin structure lacked, as did the adjacent bridge carbon, C_{20}. Swiss and British investigators therefore raised the intriguing possibility that the "missing" methine bridge in B_{12} along with its adjacent methyl group was eliminated as a two-carbon unit, perhaps in the form of acetic acid (Fig. 3.8). It was an ingenious hypothesis, published simultaneously in the same journal by the two sets of researchers and promptly proved (Eschenmoser et al., 1981; Arigoni et al., 1981; Battersby et al., 1981). This extraordinary mechanism has so far remained unique, unforeseen and only rationalized *post factum*.

Methylation of the porphyrin carbon in question (C_{20}) precedes its elimination in the form of acetic acid, instead of direct removal of a one-carbon compound. Why this roundabout mechanism? Perhaps only an oxidative mechanism not available to the anaerobic bacterial cell would be competent to perform a direct C_1 elimination.

Ribonucleotide Reductase

In recent years compelling arguments have been presented for the existence of an "RNA world" prior to the advent of DNA. One of the prerequisites for this evolutionary sequence would be a mechanism for converting ribose or a ribose derivative to deoxyribose—the replacement of a pentose hydroxyl by hydrogen, formally a reductive process (Fig. 3.10).

In fact, two distinct ribose nucleotide reductases are known (Stubbe, 1989). Some cells, such as *E. coli,* may contain both. *E. coli* is one of the relatively rare organisms known as facultative bacteria that are capable of leading two different lifestyles, one aerobic and the other anaerobic, albeit requiring different carbon sources for growth in the two instances. When *E. coli* cultures are exposed to air, a tyrosine residue in the reductase is oxidized with the aid of iron to a free radical, a necessary inter-

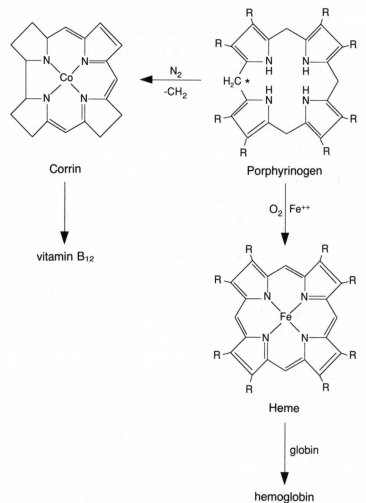

Fig. 3.8 The structure of the tetracyclic (four pyrrole rings) porphyrinogen; anaerobic conversion to corrins and aerobic conversion to heme. Only the skeletal porphyrinogen and corrin structures are shown. Peripheral methyl (CH_3), acetate ($-CH_2COOH$), and vinyl ($CH=CH_2$) groups are omitted.

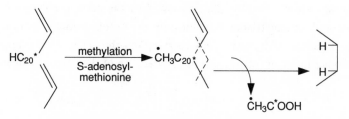

Fig. 3.9 A mechanistic detail for eliminating a methine bridge carbon from porphyrinogen to the corrin ring of vitamin B_{12}.

mediate in the ribose-deoxyribose conversion. In anaerobic *E. coli* cultures, a functionally identical enzyme not only operates in the complete absence of air but is actually poisoned by oxygen.

In the anaerobic mode, *E. coli* produces a glycine free radical instead of the tyrosine radical in the enzyme's polypeptide chain, one of the several differences between the complex mechanisms employed by aerobic and anaerobic *E. coli* (P. Reichard (1993), J. Biol. Chem. **268**, 8353). Yet the overall result is the same, reduction of ribose to deoxyribose units. Few would argue that this key reaction did not play a signal role in the evolution of organisms.

The mechanistic details of the two ribonucleotide reductases have been greatly simplified here and may be inaccurate save for the main point, the existence of aerobic and anaerobic mechanisms. Again, as emphasized earlier, a presumably primitive anaerobic mechanism has been *replaced* in evolution with the advent of atmospheric oxygen. The products of both

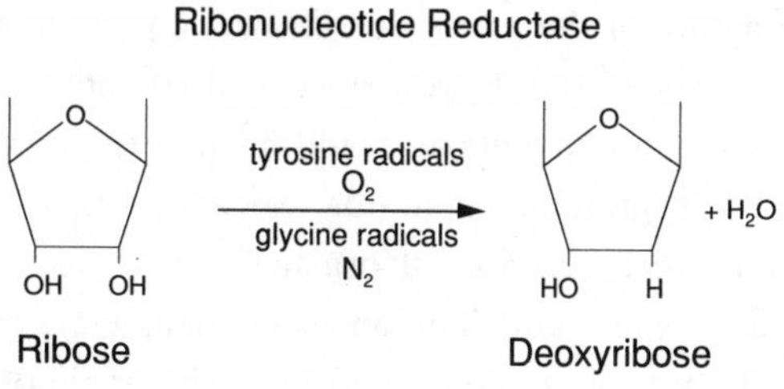

Ribonucleotide Reductase

Fig. 3.10 Aerobic and anaerobic conversion of the nucleic acid sugar ribose to deoxyribose.

processes are chemically identical. As far as is known, eukaryotic cells, yeast, and animal tissues contain only the aerobic ribonucleotide (iron-tyrosyl radical) reductase. Cells carrying out the anaerobic version failed to survive in eukaryotes, presumably because of their oxygen sensitivity.

Oxygen and Organelles

So far this discussion has emphasized evolutionary innovations of enzyme mechanisms in the wake of the appearance of atmospheric oxygen. We know much less about the corresponding or accompanying evolutionary changes at the morphological level. Was oxygen also the driving force for the prokaryotic-eukaryotic transformation, the genesis of various organelles, the nuclear membrane, the endoplasmic reticulum, and the mitochondria?

Widely held is the view that the mitochondria (organelles converting fuel to ATP) evolved by a marriage of some aerobic respiring bacterium with a "primitive eukaryote" (discussed by Whatley et al., 1979). The two precursor cells entered a symbiotic relationship, one engulfing the other. Thus the mitochondria became the power plant of all modern eukaryotic cells. What comes close to proving the point is the exceptional lifestyle of the amoeba *Pelomyxa palustris*. This otherwise eukaryotic organism, so far exceptional, lacks mitochondria and instead lives in a permanent symbiotic relationship with aerobic bacteria.* One cannot quarrel with the ingenious hypothesis that eukaryotic cells started out as endosymbiotic organisms, a concept supported by genetic evidence. What remains a puzzle is the nature of the primitive eukaryotic partner: no one has specified the properties of this hypothetical cell. It must have been an anaerobe; otherwise, no advantage would have been gained by acquisition of the bacterial respiratory apparatus. Second, was it the primitive eukaryotic partner that contributed organelles such as the nuclear membrane, the endoplasmic reticulum, and other characters present in modern eukaryotes but absent in prokaryotes? Or is the origin of these organelles

*The same type of union is thought to be responsible for the genesis of the chloroplast.

a later event, subsequent to the merger that resulted in the endosymbiotic relationship?

I am aware of only one piece of experimental evidence that could conceivably bear on the subject of oxygen and the ancestry of organellar evolution. Yeast, a fermentative and microaerophilic organism, will grow in the total absence of oxygen (with supplementation of certain lipids). It contains all the typical eukaryotic organellar characters, with one exception. In such cells normal, functional mitochondria are replaced by degenerate forms, so-called promitochondria. These incomplete organelles lack only a few proteins, notably some components of the cytochrome oxidase system. Admission of oxygen to the anaerobically grown yeast cells promptly converts the "pro" form to functionally competent mitochondria. One can reason therefore that the genesis of other organelles may not be oxygen dependent, unless the lipids needed for anaerobic growth are essential for the formation of specialized membranes. The synthesis of these lipids themselves requires oxygen. We are left with a circular argument that offers no escape from the dilemma.

Let me close this section on prokaryotic-eukaryotic descent with a quotation from Roger Stanier, who addressed a seminar I once attended: "Evolution is the kind of exercise for contemporary scientists comparable to mythology, the preoccupation of medieval philosophers. It is addictive, much like chewing peanuts."

Oxygen Toxicity

Along with the benefits gained from the advent of oxidative metabolism, a certain price had to be paid. We have already seen that oxygen is toxic to certain anaerobic bacteria. To animals, including humans, oxygen may be harmful for different reasons. Certain lipid molecules, notably the multiply unsaturated fatty acids (linoleic, linolenic, and arachidonic acids) are prone to air oxidation. They are essential but fragile molecules. When exposed to air, they break apart into reactive fragments, which in turn may inactivate vital proteins or nucleic acids.

One of the cholesterol-carrying blood plasma proteins known as LDL, or low-density lipoprotein, is ordinarily rich in polyunsaturated fatty acids.

There is evidence that oxidative modifications of LDL render animals more susceptible to atherosclerotic lesions. Experiments with rabbits have been designed to lower the polyunsaturated fatty acid content of LDL by a dietary regimen containing largely oleic acid in the form of sunflower oil, which is much less susceptible to oxidation than fats containing the more unsaturated fatty acids. The LDL isolated from the plasma of rabbits receiving sunflower oil indeed contained lesser amounts of oxidized fatty acid than did LDL samples from animals maintained on laboratory chow. In these experiments sunfower oil significantly reduced the lipid lesions in the rabbit arteries as well.

Because of the widely publicized Eskimo experience—cardiovascular diseases are rare in Greenland—fish oils have become popular dietary supplements in recent years. Yet warning signals are beginning to appear. Fish oils are one of the richest sources of $(n-3)$ polyunsaturated fatty acids, molecules highly vulnerable to oxidative damage. In order to counteract these potentially damaging effects, fish oil adherents are now advised to supplement their diets with so-called antioxidants as a countermeasure. Antioxidant chemicals have long been in use for food preservation. One of them—vitamin E, a "natural" antioxidant—may act as a defense mechanism preventing or correcting oxidative damage. Vitamin E is one of the lipid-soluble vitamins having ready access to the membrane-associated lipids rich in polyunsaturated acids. Nutritionists have given qualified approval to the use of vitamin E as an antioxidant. Nevertheless, it seems doubtful that humans consuming well-balanced diets need such supplements.

Attack and Defense

A variety of agents—chemical, physical, and biological—convert oxygen into toxic forms: hydrogen peroxide (H_2O_2), ozone (O_3), superoxide anion, and singlet oxygen (Fig. 3.11). All of these oxygen species are more powerful oxidizing agents than the oxygen we breathe. This fact may be contrary to the body's spontaneous response: we breathe more deeply in a poorly ventilated room or in the rarefied air at high altitudes. Thus, the adoption of respiration, the aerobic lifestyle, has its inherent risks.

1) 4 cytochrome c + $4H^+ + O_2 \longrightarrow 2H_2O$

2) $O_2 \xrightarrow{\quad e^- \quad} O_2^- \cdot$
    superoxide
    anion

3) $O_2^- \cdot + O_2^- \cdot \xrightarrow[\text{superoxide dismutase}]{2H^+} H_2O_2 + O_2$

4) $H_2O_2 + H_2O_2 \xrightarrow{\text{catalase}} 2H_2O + O_2$

Fig. 3.11 Reactions for eliminating toxic forms of oxygen.

In the terminal events of respiration, reduced cytochromes transfer four of their electrons essentially simultaneously to form two molecules of water (1, Fig. 3.11). However, in what may be called chemical side reactions—intrinsic properties of the O_2 molecule—electrons can be transferred singly and sequentially, with potentially dire consequences. Addition of a single electron to oxygen generates a free radical, known as superoxide anion (2, Fig. 3.11), in turn a highly reactive oxygen species that creates havoc with a number of biomolecules. Ionizing radiation generates such free radicals capable, for example, of breaking DNA strands.

Presumably, ever since oxygen appeared in the atmosphere, Nature has been inventing countermeasures to render superoxide anion less harmful, if not entirely innocuous. One of the two protective enzymes, known as superoxide dismutase (Fridovich, 1986), converts superoxide anion to H_2O_2 (3, Fig. 3.11). In turn, the ubiquitous enzyme catalase (4, Fig. 3.11), present in all respiring cells, eukaryotic and prokaryotic, disposes of H_2O_2, converting it to water and oxygen.

In animal tissues superoxide dismutases are metallo-enzymes and are of two types. Cytoplasmic or soluble dismutases contain both copper and zinc, while in the mitochondrial enzyme the activating metal is manganese. The dismutase of aerobic bacteria is similar and related to the eukaryotic

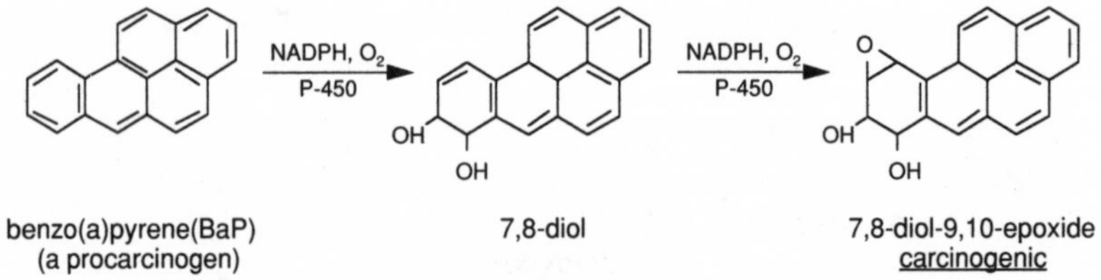

Fig. 3.12 Oxidative conversion of the procarcinogen benzopyrene to
a carcinogen.

manganese variety; it also shows DNA sequence homology with the mammalian enzyme. Once again, genetic evidence is consistent with an endosymbiotic origin of eukaryotic cells from an aerobic bacterium and a primitive eukaryotic partner.

Xenobiotics

The world's population is exposed—both inadvertently and no longer unaware—to an ever-growing variety of chemical substances that are alien (not made by the body), known as xenobiotics (*xeno,* foreign). They include environmental poisons and the synthetic molecules produced by the chemical and pharmaceutical industries. To varying degrees, they are harmful either as they are or when converted in the body to toxic substances. The ultimate reason for toxicity is Nature's imperfection. The animal body can handle xenobiotics because enzymes and receptors interact with foreign as well as physiological molecules. The phenomenon of drug adaptation or tolerance is an example. When given over extended periods, phenobarbital "induces" enzymes in the body that chemically modify the drug and diminish its sedative potency.

In many instances, carcinogens are not tumor producing per se but only after oxidative tissue transformation into reactive metabolites. A notorious example is the hydrocarbon benzopyrene (BaP), one of the noxious substances in tobacco smoke (Fig. 3.12). When inhaled, BaP is metabolized to toxic products—for instance, in the nasal cavity of the Syrian hamster, in the isolated, perfused lung of rats, and in the tracheal and bronchial tissues of several species.

As shown in Fig. 3.12,* the first step leads to the introduction of two OH groups into the hydrocarbon, to form a 7,8-diol, followed by entry of an additional oxygen in the form of the 9,10-epoxide. This second oxygenase product, the epoxide, is the carcinogen that reacts irreversibly with a segment of DNA. The inhaled benzopyrene is the *pro*carcinogen. Let me emphasize once again that the body's adverse response to foreign substances results from the fact that enzyme or receptor specificities are not absolute.

Bibliography

1. J. R. Nursall (1959), Oxygen as a prerequisite to the origin of Metazoa, *Nature* 183, 1170–71.

2. H. Goldfine and K. Bloch (1963), *Control mechanisms in respiration and fermentation,* pp. 81–103. New York, Ronald Press.

3. N. Morita et al. (1992), Both the anaerobic pathway and aerobic desaturation are involved in the synthesis of unsaturated fatty acid in *Vibrio,* sp. strain ADE-1, *FEBS LETT.* 297, 9–12.

4. A. E. Garrod (1909), *Inborn error of metabolism,* London, Frowde, Hoddes and Stonghton.

5. M. S. Shorb (1948), Activity of vitamin B_{12} for the growth of *Lactobacillum lactis, Science* 107, 397.

6. D. Hodgkin et al. (1954), Structure of vitamin B_{12}, *Nature* 176, 325–238.

7. A. Eschenmoser et al. (1981), ring contraction of hydroporphinoid to corronoid complexes, *Proc. Nat. Acad. Sci.* (USA) 78, 16-19.

8. D. Arigoni et al. (1981), Biosynthesis of vitamin B_{12}, *Proc. Nat. Acad. Sci.* (USA) 78, 11–12.

9. A. R. Battersby et al. (1981), Biosynthesis of vitamin B_{12}, *Proc. Nat. Acad. Sci.* (USA) 78, 13–14.

*The "P-450" in Fig. 3.11 refers to specialized cytochromes that catalyze a variety of oxygenase reactions, including those responsible for the formation of sex hormones. It was in this way that P-450 was discovered. The "P" stands for porphyrin, and the "450" for the wavelength of maximal light absorption of the CO-complex of the porphyrin.

10. J. Stubbe (1989), Protein radical involvement in biological catalysis? *Ann. Rev. Biochem.* 58, 257–285.

11. J. M. Whatley, P. John, and F. R. Whatley (1979), The establishment of mitochondria and chlorophlasts, *Proc. Roy. Soc. Long. (Biol)* 204, 165–187.

12. I. Fridovich (1986), Superoxide dismutases, *Adv. Enzymol.* 58, 61–97.

4

The Importance of Being Contaminated

———

Unsuspected reagent contamination was responsible for

the discovery of a number of biomolecules. For instance,

three important biochemical processes were discovered

because of impure ATP. Other examples are contamination

of pancreatic insulin with the hormone glucagon, and

chemical contamination of an enzyme substrate that led

to the discovery of "enzyme suicide."

Biochemical experimentation has been enormously expedited by the growing availability of numerous reagents from commercial sources. Scientists of my generation remember the days when common biochemicals such as ATP had to be prepared from rabbit muscle in one's own laboratory, not a difficult but a time-consuming task. Few, if any, enzymes could be bought unless they were also used in industry. The purity of homemade or commercial chemicals was not a major concern as long as the reagent had the desired activity. Older analytical techniques were not sufficiently sensitive to detect minor impurities.

Years ago the German company Schering-Kahlbaum sold chemicals under the label "purissimum." Theirs were indeed the finest chemicals available at the time. Yet "pure" is a qualitative term, no longer acceptable unless upper limits of contamination are specified. For most experimental purposes trace impurities are innocuous and do not affect the results. Still, we now know that minute amounts of biomolecules exert biological activity unrelated to statements on the reagent's label.

Reagent contamination may arise from two causes, intrinisic instability (deterioration of the desired chemical) and/or inadequate procedures for separating the desired compound from unknown accompanying substances.

In the instances to be described here, contamination due to unsuspected bioactive molecules proved to be instrumental in major discoveries. Needless to say, such discoveries were of the serendipitous kind. I suspect that a number of similar situations were never made public, because investigators tend to take more pride in discoveries that originated in their minds than in those caused by accidental events.

Oxysterols

Cholesterol when freshly isolated from animal tissues is a white, crystalline, odorless material. On standing at room temperature it turns quite rapidly into an evil-smelling yellowish substance. In order to prevent this deterioration, cholesterol needs to be kept frozen in airtight bottles. Evidently oxygen and presumably light cause cholesterol molecules to decay. As early as 1940, a rapid air oxidation of cholesterol in aqueous suspension to 7-hydroxycholesterol (Fig. 4.1) was noted. This chemical instability became a serious problem when evidence began to mount that cholesterol synthesis in the animal body is inhibited by dietary cholesterol itself, a phenomenon known as negative feedback inhibition. (In other words, a metabolic end product can regulate its own synthesis.) Later findings identified the cholesterol-sensitive feedback site at the level of hydroxymethylglutaryl coenzyme A (HMG-CoA) reductase (HMG:β-hydroxy-β-methylglutaryl), one of the numerous enzymes participating in cholesterol biosynthesis.

Andrew Kandutsch, interested in this regulatory phenomenon, studied

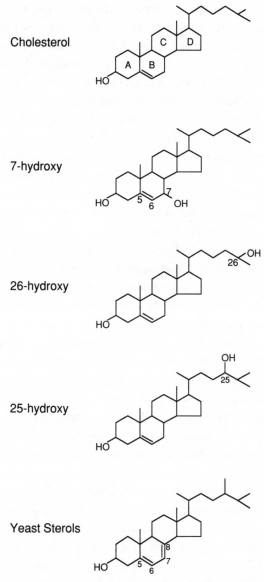

Fig. 4.1 Hydroxylated sterols inhibiting cholesterol biosynthesis
and HMG-CoA reductase activity in mouse liver cells.

sterol inhibition of the reductase enzyme in cultures of mouse liver cells. He found the extent of enzyme repression to be variable, depending on the quality or age of the commercial cholesterol he added to the culture medium. Suspecting the presence of trace contaminants, he purified a batch of cholesterol by repeated crystallizations and, more effectively, by gas chromatography. Cholesterol thus purified, when added to cultured mouse liver cells, inhibited cholesterol synthesis and repressed the activity of the rate-limiting HMG-CoA reductase only slightly. Older, impure sterol samples, however, impaired these processes substantially. There was good reason to suspect that the contaminating cholesterol companions were oxidized derivatives of cholesterol.

At least half a dozen such contaminants were identified. They contained additional OH groups at various positions of the sterol ring and side chain (Fig. 4.1). When tested individually in various lines of cultured animal and human cells, several of these oxidation products affected cholesterol bio-synthesis to a greater extent than purified cholesterol. Thus the question arose whether cholesterol itself regulated its own synthesis, or whether a product of oxidative cholesterol metabolism was responsible, or whether both were involved in the control. The issue remains unsettled today.

Observations that small amounts of oxycholesterols are found normally in tissues further complicates this issue. Are they in vivo "artifacts"? At least two of the hydroxy derivatives definitely are not in this category. For example, 7-hydroxycholesterol is the first intermediate on the pathway of cholesterol to bile acids, while 20-hydroxycholesterol serves as a normal precursor of steroid hormones. But, another, 25-hydroxycholesterol, is apparently an artifact.

In attempts to distinguish the effects of cholesterol from those of its oxidation products, Kandutsch discovered a protein that strongly attaches to various oxysterols but is distinct from a protein that binds cholesterol itself. It is well known that binding to carrier proteins is involved in receptor-mediated gene expression of, say, steroid hormones.

A hypothetical scheme proposed by Kandutsch, "consistent with reg-ulation of cholesterol synthesis in different tissues" (Taylor and Kan-dutsch, *J. Chem. Phys. Lipids,* 1985), embodies three postulates: (1) in replicating cells that do not catabolize cholesterol but employ cholesterol as such for membrane function, an oxidized precursor sterol may regulate the activity of HMG-CoA reductase; (2) in liver cells cholesterol synthesis destined for bile acid production may be regulated by bile acid precursors such as 7-hydroxycholesterol; and (3) in endocrine glands producing the cholesterol-derived male and female sex hormones, 20α-hydroxycholesterol may regulate the synthesis of cholesterol destined for hormone production. As Kandutsch admits, his ad hoc hypothesis needs verification, but there can be no doubt that his accidental discovery of oxysterols emphasizes the complexity of the mechanism by which cholesterol and its potential metabolites may regulate cholesterol homeostasis. These results complement rather than contradict the model and brilliant research of Brown and Goldstein beginning in 1975, on the general regulatory role of plasma LDL in cholesterol synthesis.

As pointed out at the beginning of this chapter, cholesterol's sensitivity to autoxidation is ultimately responsible for the discovery of what may be an alternative mechanism for the regulation of cholesterol biosynthesis. One can argue, for what it is worth, that in vivo reduction of cholesterol's 5,6-sterol double bond (Fig. 4.1) to the more stable cholestanol in the body would provide protection against oxidation, but Nature has not chosen to follow this route. Presumably the retention of this 5,6 double bond is essential for the further metabolism to bile acids and steroid hormones.

In fact, a sterol that is even more oxygen sensitive is 7-dehydrocholesterol containing two double bonds (in conjugation) in the same ring (5,6 and 7,8; see Fig. 4.1). This intermediate in sterol biosynthesis is reduced partially to cholesterol in animal tissues,* while in yeast the sterol diene system is retained. In the yeast cell, normally a fermentative organism, the intracellular oxygen tension is much lower than in the body fluids of animals, and therefore oxidative damage to sterols in yeast will be minimal. This attempt to explain why animal sterols contain one and yeast

*The exception is 7-dehydrocholesterol in skin. Here the sterol diene exposed to oxygen and ultraviolet light is converted oxidatively to vitamin D or cholecalciferol.

sterols contain two double bonds in ring B is pure speculation. For whatever reason, the diene structure seems to be optimal for yeast.

Vanadium and ATPase

Your own failure to repeat an experiment and that of a colleague may have different causes. As we shall see in Chapter 11, one strain of rats will respond to insulin, but another will not. Or, as in the preceding example, autoxidized cholesterol inhibits cholesterol biosynthesis, whereas cholesterol freed of contaminating oxysterols will not show this inhibitory effect under certain conditions.

Probably the most widely used reagent in biochemistry is ATP. It has been available commercially for several decades. Yet evidence for contaminants that drastically affect the biological activity of ATP samples has come to light only in more recent years. Three examples will be cited.

In the mid-1970s several laboratories studies the ATP-driven transport system for sodium and potassium ions, an enzyme system known as the Na^+-K^+ pump. The function of the enzyme is to maintain high potassium and low sodium levels inside the cell, relative to their external (extracellular) concentrations. It is ATP hydrolysis that provides the energy. Nearly one-third of the body's ATP is consumed in this ion pumping. From time to time several laboratories reported variable results, some samples showing ATP stimulation and others, paradoxically, inhibition of the ion pump. Lew Cantley and Guido Guidotti, my colleagues in the biochemistry department at Harvard, examined these inconsistencies. They suspected that ATP preparations, when inhibitory, were contaminated. Standard purification procedures failed to turn "bad" ATP into "good" ATP. At that time the Sigma Company supplied most of the samples of ATP for the biochemical community. As late as 1977, Sigma's customers apparently did not or had no reason to complain about the quality of ATP. Sigma in fact supplied two different ATP preparations, one isolated from horse muscle (Sigma grade ATP), and another, labeled Sigma grade I, derived from yeast. The former inhibited the $(Na^+$-$K^+)$ ATPase, and the latter, acting properly, promoted the ion pump.

Eventually Cantley and Guidotti succeeded in separating the muscle-derived ATP from the suspected "impurity" by high-powered chromatography. Two fractions were obtained, one that behaved normally, was indistinguishable from yeast ATP, and stimulated the ion pump. The second fraction contained a very powerful ion pump inhibitor. In the contaminated muscle ATP, the inhibitor was present in extremely low concentrations: three parts per 10,000, or 0.03 percent. Chemical tests revealed that this material was stable to acid and alkali, much more so than ATP, and that it was a significantly smaller molecule. There were no indications that the inhibitory contaminant was chemically related to ATP, even though it had a very high affinity for the ATPase enzyme.

After a few months Cantley and Guidotti suspected a metal as the culprit. A sample of the purified inhibitor was sent to the laboratory of Bert Valley for metal analysis by a highly sensitive method known as microwave-induced emission spectroscopy. The results were unambiguous. While thirteen metals (among them copper, zinc, and chromium) were identified, oxides of vanadium accounted for over 90 percent of the sample's mass. Authentic vanadium oxide (V_2O_5) was secured and shown to have the same inhibitory effects on ATPase activity as the factor separated from Sigma muscle ATP.

Clearly, vanadium is an unusually potent enzyme inhibitor, but was its effect physiologically significant? Also, why should only muscle-derived ATP be contaminated and not yeast ATP, even though the commercial purification procedures were presumably the same? One wonders how much longer the discovery of vanadium and its role in animal cells would have taken if the biochemists had chosen yeast-derived ATP exclusively. For yeast, in contrast to animal tissues, contains no vanadium.

Vanadium's potency—that is, its inhibition of ATP cleavage at very low concentrations—is attributed to an unusually high affinity for the enzyme's ATP-binding site. Vanadium,* a widespread element in Nature, comprises 0.01 percent of the earth's crust and occurs in some sixty five minerals.**

*Vanadium was discovered in 1801 by Sesström, a Swede. He named the metal after Vanadis, an old Norse designation for Ireya, the goddess of love.

**Geochemists have found vanadium in petroleum, apparently tightly linked to porphyrins.

The metal is best known in the form of vanadium steel, a rust-resistant alloy.

Most animal tissues contain vanadium in significant amounts (10^{-5}–10^{-6} M), indicating that the metal, when present in the diet, is absorbed into the bloodstream. The metal is not concentrated in a specific organ or subcellular organelle. Rats, when kept on a synthetic diet lacking vanadium, suffer retardation of growth and, more specifically, deficiencies in the respiratory processes. All of the deficiency syndromes can be corrected by supplementing diets with as little as 100 parts per billion of the metal.

The key to the still-unknown function of vanadium or its oxides may be that it shares some properties with phosphate ions. The metal attaches to and competes with numerous binding sites for ATP and related phosphate compounds, especially of sugars. Also, vanadium combines readily with numerous proteins.*

According to a recent report, among its other positive effects vanadium ameliorates certain defects associated with drug-induced diabetes. One of these diabetogenic lesions is associated with an inhibition of glycogen synthesis. Vanadium will correct the inhibition in isolated rat liver cells. In this instance at least, vanadium's effects appear to be stimulatory rather than inhibitory. Still, the principal targets of this metal and whether it is a major metabolic regulator, for instance of the ion pump, remain unknown.

An exotic class of organisms not in the mainstream of biochemical research may provide a clue. They are the tunicates (or sea squirts), marine invertebrates possessing several external gelatinous coats (tunics, or mantles). Quite remarkably, these organisms concentrate vanadium

---

*Depending on oxygen tension and pH, vanadium exists in three interconvertible oxidation states, +3, +4, and +5, a factor that complicates mechanistic studies. In most biological systems ^{3}V, or Na_3VO_4, appears to predominate. Vanadium oxides also polymerize readily to dimers and tetramers. Lubert Stryer (*Biochemistry,* 3rd ed., New York, W.H. Freeman, 1988) mentions "vanadate as a transition state analogue in phosphoryl transfer reactions because it forms a pentacovalent bipyramidal structure like that of phosphate esters undergoing hydrolysis" (p. 952).

from seawater and store the metal in specialized blood cells, doing so against a concentration gradient of 10^{-8} M! Seawater contains only 10^{-8} M, and the tunicate blood cell contains 1.0 M vanadium. Although vanadium has a high affinity for oxygen, tunicates apparently do not use this metal for carrying oxygen to the tissues. It has been suggested that vanadium functions in some manner during the synthesis of the "tunic" envelope, a process specific to the tunicate family.

To the reader concerned about an adequate dietary supply of vanadium, a volume entitled *Present Knowledge in Nutrition* (1990) offers reassurance: "The failure to establish vanadium as an essential element prevents any suggestion of a vanadium requirement. Besides, foods rich in vanadium include shellfish, mushroom, parsley, dill seed, and black pepper." Presumably normal or balanced diets contain amounts sufficient to meet the body's needs.

Vanadium's biological role is not confined to animal tissues. Higher and lower plants (algae) contain the metal, which appears to be essential for nitrogen fixation in some instances. This process converts the inert atmospheric nitrogen into ammonia (fixed nitrogen). It takes place in the roots of leguminous plants with the help of symbiotic bacteria, the rhizobia. Free-living bacteria that can fix nitrogen are known as azotobacteria. The complex machinery for converting N_2 into NH_3 includes two metalloproteins, one containing iron and the other iron and the rare metal molybdenum.

The observations of one of the research groups working on this problem, the "unit of nitrogen fixation" at the University of Sussex, England, deserve special mention here because they belong to the rubric of "the importance of being contaminated." J.R. Postgate and his group worked with a genetically engineered strain of *Azotobacter chromococcum,* which grew and fixed nitrogen on a medium containing sodium tungstate but free of molybdenum. The requirement for tungstate (Na_2WO_4)* was inordinately high, leading to the suspicion that the tungstate sample was con-

---

*W stands for Wolfram (possibly Wolf's foot?), the German name for tungsten, which was discovered by the Swedish chemist K. W. Scheele in 1781. Like vanadium, it is used in specialty steels. Uniform symbols for quite a number of elements are yet to be agreed on by the science community.

taminated with another metal. A method similar to the one that revealed the presence of vanadium in ATP showed that the same metal was responsible for supporting nitrogen fixation in *A. chromococcum*. Why molybdenum is the required metal for nitrogen fixation in some *Azotobacter* species and vanadium in others remains a mystery. The two metals are functionally identical but not interchangeable—a striking example of biochemical diversity.

Another Example of ATP Contamination

Among the numerous contributions to modern biochemistry of Arthur Kornberg, there is one well known especially to lipid biochemists. In 1952, as he relates in *For the Love of Enzymes* (Harvard University Press, 1988): "My first of several affairs with membranes grew out of the mistaken notion that the synthesis of the relatively simple phospholipid molecules might be an instructive model for how the backbone of a complex nucleic acid chain is built. Synthesis of the phosphate bridge between the two 'alcohol' (OH) moieties of a phospholipid (glycerol and choline) might teach us how the phosphate connection between the alcohol-like sugars of a nucleic acid backbone was made."

Two related papers of Kornberg and Pricer emerged from that "mistaken" notion. They were important contributions to phospholipid biosynthesis, but as it turned out not relevant to nucleic acid biosynthesis. Kornberg's principal results indicated that phosphorylcholine might be incorporated as such into the membrane lipid phosphatidylcholine, and that the process required ATP as well as phosphocholine, an unexpected finding (Fig. 4.2).

Eugene Kennedy, unaware of Kornberg's experiments, asked himself an essentially similar question (1954): what is the mechanism for the formation of the phosphodiester bridge in phosphatidylcholine? One major difference between Kennedy's and Kornberg's experimental designs was that Kennedy did not *add* the required ATP from a reagent bottle as Kornberg had done, but generated it in situ by a process known as oxidative phosphorylation. This experimental detail proved to be crucial. Under Kennedy's condition only free choline, not phosphocholine, was

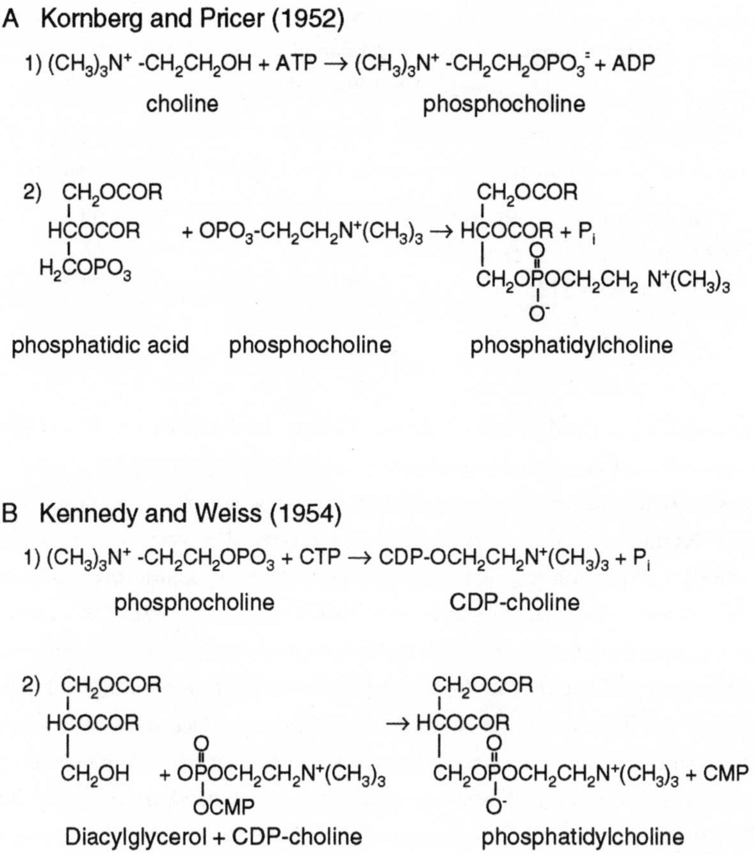

A Kornberg and Pricer (1952)

1) $(CH_3)_3N^+ -CH_2CH_2OH + ATP \rightarrow (CH_3)_3N^+ -CH_2CH_2OPO_3^{=} + ADP$

       choline   phosphocholine

2)
$$
\begin{array}{l}
CH_2OCOR \\
| \\
HCOCOR \\
| \\
H_2COPO_3
\end{array}
+ OPO_3\text{-}CH_2CH_2N^+(CH_3)_3 \rightarrow
\begin{array}{l}
CH_2OCOR \\
| \\
HCOCOR + P_i \\
|\quad O \\
CH_2OPOCH_2CH_2\ N^+(CH_3)_3 \\
\quad\quad | \\
\quad\quad O^-
\end{array}
$$

phosphatidic acid   phosphocholine   phosphatidylcholine

B Kennedy and Weiss (1954)

1) $(CH_3)_3N^+ -CH_2CH_2OPO_3 + CTP \rightarrow CDP\text{-}OCH_2CH_2N^+(CH_3)_3 + P_i$

        phosphocholine   CDP-choline

2)
$$
\begin{array}{l}
CH_2OCOR \\
| \\
HCOCOR \\
| \\
CH_2OH
\end{array}
+
\begin{array}{l}
O \\
OPOCH_2CH_2N^+(CH_3)_3 \\
| \\
OCMP
\end{array}
\rightarrow
\begin{array}{l}
CH_2OCOR \\
| \\
HCOCOR \\
|\quad O \\
CH_2OPOCH_2CH_2N^+(CH_3)_3 + CMP \\
\quad\quad | \\
\quad\quad O^-
\end{array}
$$

Diacylglycerol + CDP-choline   phosphatidylcholine

Fig. 4.2 Two modes of phosphatidylcholine biosynthesis, discovered
independently.

Table 4.1

Cofactors needed for the conversion of phophorylcholine to phosphatidylcholine (PC)

Cofactor added	PC synthesized
ATP, Pabst lot 116, amorphous	560
ATP, Pabst lot 112, crystalline	20
ATP, Pabst lot 112, crystalline, + CTP	1,677

Source: Kennedy, 1962.

converted to a lipid form. Did this finding cast doubt on the role of phosphocholine as an intermediate?

For a while the problem remained unresolved. When Samuel Weiss joined Kennedy's laboratory in 1954, he repeated Kornberg's as well as Kennedy's experiments. Both were confirmed. At this juncture, Kennedy (1962) recalls, "we then decided to confront the anomaly that the conversion of phosphocholine to phosphatidylcholine required *ATP from the bottle,* while ATP generated [in situ] via oxidative phosphorylation was ineffective." The suspicion that one of the two ATP samples, but not the other, contained a highly active impurity needed for phosphocholine synthesis was borne out. Kornberg and Pricer had used *amorphous* ATP taken from a bottle of commercial (Pabst) ATP. Kennedy had access to *crystalline* ATP (100 percent pure?) in 1954, when it became first available from the Pabst Laboratories. Both ATP samples were tested, with the results shown in Table 4.1.

Kennedy tested a number of nucleoside triphosphates (GTP, ITP, and CTP) because these nucleotides were likely contaminants of ATP. Only CTP—in combination with ATP (the inactive lot of crystalline ATP)—stimulated the conversion of phosphocholine, and it did so even more effectively than amorphous ATP alone (Table 4.1.). The CTP, or cytidine triphosphate, obviously contaminated amorphous ATP but was removed on crystallization. Inspired reasoning that the contaminant might have properties closely similar to that of ATP led to success.

Kornberg's contribution was having identified the first step in phosphatidylcholine synthesis, the phosphorylation of choline to phosphocholine

1) choline + ATP $\xrightarrow{\text{choline kinase}}$ phosphocholine

2) phosphocholine + CTP $\longrightarrow$ cytidinediphosphate-choline + PP_i (CDP choline)

Fig. 4.3 Phospholipid synthesis from choline involves two activation reactions, one requiring ATP and the other CTP.

with the aid of an enzyme called choline kinase. Unaware that his amorphous ATP also contained CTP, he had every reason to believe that phosphocholine was the "active" choline donor for phospholipid synthesis. Kennedy's good fortune in having access to crystalline ATP enabled him to show that phospholipid synthesis from choline involved *two* activation reactions, one requiring ATP and the second CTP (Fig. 4.3). These landmark investigations eventually clarified the enzymatic pathways to all six phospholipid constituents of animal membranes.

Why Nature has chosen cytosine-containing CTP as the specific cofactor for phospholipid biosynthesis we do not know. Why not the nucleoside triphosphates that contain other nucleic acid bases such as adenine (ATP), guanine (GTP), or uracil (UTP)? Also, we can only raise, but not answer, the question, Why is the uracil derivative UDPG the activated form of glucose in glycogen synthesis? And finally, why is GTP central to hormone-receptor interactions?

I learned only accidentally of another example of contaminated ATP, in this case with GTP, guanosine triphosphate. Readers of the recent biochemical literature on G-proteins will be aware of its importance.

In 1971 the laboratory of Martin Rodbell published research entitled "The Glucagon-Sensitive Adenylic Cyclase System in Plasma Membranes of Rat Liver, Effects of Guanylnucleotides" (*J. Biol. Chem.* **246**, 1872–77). It dealt with the role of ATP in the binding of the hyperglycemic hormone glucagon to the plasma cell receptor. Different results were obtained when commercial ATP "from the bottle" was compared with ATP generated in the test system by oxidative phosphorylation. Recalling the earlier experiences of Arthur Kornberg and Eugene Kennedy with ATP of diverse origins, Rodbell suspected that the commercial ATP might not be free from trace amounts of other nucleotides. Tests with GTP, ITP, and UTP showed that pure GTP, and only GTP, gave the same qualitative response as ATP on hormone binding but at concentrations a thousandfold less. This finding eventually led to the discovery of the large family of G-protein molecules that serve in diverse signaling reactions initiated by hormones and mediated by membrane receptors. The process known as signal transduction is ultimately expressed by enzymatic events in the cell's interior. The G-proteins turned out to be key players in various hormone-controlled events, and finding them was one of the biochemists' quests in the search for the Holy Grail. (Gilman, *Ann. Rev. Biochem.* **56** [1987], 615).

It is intriguing that the four base moieties of RNA (adenine, guanine, cytosine, and uracil) in the form of their nucleoside triphosphates also participate in the major and probably universal metabolic processes. Thus each letter of the genetic code has the additional role of specifying a host of metabolic pathways and control events. What remains to be rationalized is the specificity of the letters in relation to their chemical structure.

Glucagon

The beginnings of hormone research date to 1889, when the surgeon J.R. Mehring daringly extirpated the pancreases of experimental animals.* He demonstrated that this surgical removal rendered the animals diabetic. The disease *Diabetes mellitus* owes its name to the diabetic's excretion of "honey-sweet" urine.** Diabetes is diagnosed and characterized by abnormally high levels of blood and urinary glucose, which return to normal on injection of extracts of the pancreas from a variety of mammalian species. The isolation of insulin, the active pancreatic principle, by F.G. Banting, C. Best, and D.A. Scott in 1921 constitutes one of the most remarkable and widely known achievements of early biochemistry. Still, research on the molecular mode of insulin's action continues to this day, illustrating that intimate understanding at the molecular level is not

*Surgical removal of various internal or endocrine organs—the thyroid, parathyroid, adrenal glands, testes, pituitary—led to the discovery of hormones (excitatory molecules).

**The taste was the early diagnostic test for the disease.

The Importance of Being Contaminated

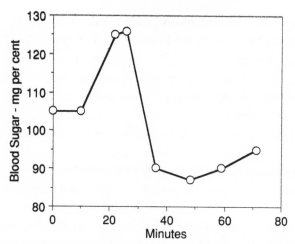

Fig. 4.4 Short-term and long-term effects of glucagon-contaminated insulin on blood sugar levels in rabbits.

essential for a successful therapy. For the diagnosis of diabetes, the standard clinical procedure measures blood sugar in the fasting state. Injection of insulin is prescribed when glucose levels exceed 100–200 mg percent.

Soon after insulin became commercially available in 1920, it was noted that the insulin-induced lowering of blood sugar was time dependent. The peak response occurred one to two hours after insulin injection. Curiously, it was sometimes preceded by hyperglycemia, a transient rise in blood sugar (see Fig. 4.4). Even more puzzling, insulin samples from the same source but different batches failed to give uniform results. With some insulins the early rise of blood sugar levels would occur, and with others the hyperglycemic effect was small or entirely absent. Was it reasonable to ascribe these opposing effects to one and the same pancreatic hormone?

As early as 1923 J. Murlin and colleagues at the University of Rochester, working with aqueous pancreatic extracts of pig and ox, came to the surprising and still-to-be-documented conclusion that "there are two substances in these extracts, one lowering the blood sugar, the other raising the blood sugar of both normal and pancreatectomized animals."* The

*It is of interest that the first edition of *General Biochemistry*, the 1953 text by J. Fruton and S. Simmonds (New York, John Wiley), mentions a "hyper-

Murlin group used the extraction procedure described in the first paper by Banting and Best (extraction of the tissue with ethanolic HCI). Murlin and coworkers were also the first to name the hyperglycemic principle "glucagon." Still, there was no proof that glucagon was a separate entity. Though unlikely, the hyperglycemic effect could conceivably be an intrinsic property of insulin.

Since all of the early insulin preparations were amorphous, there was no test for their purity. A clear-cut resolution of the issue appeared to be at hand when John Jacob Abel at Johns Hopkins University succeeded in crystallizing insulin in 1926. His insulin crystals did *not* produce the initial hyperglycemia under any conditions, strongly supporting the contamination hypothesis. It seemed evident that Abel's crystallization procedure— a classic test for purity—had removed the hyperglycemic factor.

What muddied the waters was the experience with another insulin sample, this one crystallized in the presence of zinc salts by Scott (1934). In contrast to Abel's crystals, the Scott preparation was not free of glucagon, the hyperglycemic factor. In 1939 Danish investigators (Lundsgaard and colleagues) compared the two crystalline insulin preparations under identical conditions, noting that they behaved differently in a liver perfusion test suitable for quantitative assay. The crystalline insulin sample supplied by Scott caused an initial rise in glucose production, whereas Abel's insulin crystals did not. Throughout the 1940s this discrepancy persisted.

Another attempt that promised resolution of the problem was undertaken by Christian de Duve. He found that a crystalline insulin sample provided by an American supplier (presumably prepared by Scott's method, but not documented) contained glucagon, while a Danish (Novo) insulin was free of the hyperglycemic factor. Both the American and Danish insulins were crystalline. Unfortunately, the literature does not disclose details of the commercial procedures for purifying insulin and whether they differed substantially.*

––––––––––––––––

glycemic factor" that frequently accompanies insulin preparations; the second edition in 1958 refers to this factor as glucagon and describes its crystallization and complete amino acid composition—another example of the many startling developments of the mid-1950s.

*The respective patents should provide an answer, but I have not examined

Audy and Keiley (1952) carefully examined insulin's glycogenolytic effects in rat liver slices (another test for glucagon). Eli Lilly insulin prepared in 1947 served as the reference standard. Insulin samples from five different British drug companies were tested. All contained glucagon, but less than the Lilly standard. Distressingly, different batches from the same supplier varied substantially in glucagon content, suggesting that, at that time, the purification procedures were inadequately standardized. Only the Danish (Novo) insulin gave uniform results, confirming the earlier findings of de Duve. Audy and Keiley mention that a later sample of Lilly insulin (1948) was also free of glycogenolytic factor. Evidently this manufacturer modified the purification procedure at some point, presumably between 1947 and 1948.

To summarize the discussion so far, a critical observation, apparently made in the 1920s, was the biphasic effect on blood glucose levels that followed injection of some samples of insulin. In both humans and animals the pancreatic hormone(s) cause a transient hyperglycemia, reaching a maximum about twenty minutes after injection. This effect is then overshadowed by a rapid fall of blood sugar to levels far below normal. The end result is hypoglycemia (Fig. 4.4), the therapeutically important event.

Insulin specimens, whether amorphous or crystalline, may or may not contain the hyperglycemic factor, depending on the commercial source of the hormone. Whether earlier standardization of procedures for preparing insulin might have led sooner to the discovery of glucagon is a moot question.

Properties Distinguishing Glucagon from Insulin

Soon after the transient hyperglycemic effects of pancreatic extracts became known, evidence accumulated that the two active principles responsible for regulating carbohydrate metabolism were of different origin. The pancreas contains two types of cells: β, or islet cells (islets of Lan-

---

them. Some details of the Danish patent were kindly communicated to me by Ole Hollan of Novo Nordisk.

Table 4.2
Properties of pancreatic hormones

Occurrence	Insulin β cells	Glucagon (HGF) α cells
Molecular weight	12,000	3,550
Cysteine	+	−
Effect of cobalt	−	+
Heat resistance	−	+
Optimal activity after injection	30–60 minutes	5–10 minutes
Alloxan-induced diabetes	↓	−
Glycogen breakdown	−	↓
Zinc needed for activity	+	−

gerhans), and α, or acinar cells. The β cells produce insulin; the α cells, glucagon.

A number of properties differentiate the two hormones (Table 4.2.). A diabetogenic drug known as alloxan selectively destroys the insulin-producing β cells, without impairing the glucagon activity residing in α cells. Conversely, administration of cobalt salts impairs the production of glucagon but leaves insulin production intact. In retrospect, insulin and glucagon could have been obtained individually by separating α from β cells at an early date, but apparently this procedure was impractical or not feasible technically.

The distinguishing chemical properties of glucagon include resistance to boiling in alkaline solutions, treatment with cysteine, and destruction of glucagon-producing α cells by cobalt salts. Eventually, after major advances in protein separation techniques, O. K. Behrens and W. W. Bromer in the Lilly laboratories succeeded in preparing pure (homogeneous) glucagon in crystalline form (1958). The protein had a molecular weight of 3,550, less than one-third that of insulin (12,000). Likewise, the amino acid sequence of the secreted hormone showed no resemblance to the amino acid composition or sequence of insulin. Notably, in contrast to insulin, glucagon contained no cysteine.

Pharmacologically, glucagon is apparently used only in special circumstances to combat severe hypoglycemia in diabetic patients who have taken an overdose of insulin. It is the drug of choice in emergency situations for comatose patients unable to take oral glucose solution. Like insulin, glucagon must be administered parenterally, since both are proteins susceptible to hydrolytic degradation in the intestinal tract.

As to the overall or physiological action of glucagon, it is clearly antagonistic to that of insulin. Yet at the molecular level the two hormones have separate targets. In contemporary endocrinology or pharmacology an antagonist is defined as a molecule that competes with a given hormone for binding sites to a specific receptor. The insulin receptor, now well characterized, has no affinity for glucagon. Apparently a specific binding or receptor site for glucagon alone has not yet been detected, or perhaps even sought. Also, metabolic disorders attributable to a genetic glucagon deficiency do not seem to exist.

Glucagon-contaminated insulin seems never to have been a serious issue in diabetes therapy. For the last seventy years diabetic patients have been maintained in good health throughout their lives by procedures that have not materially changed—with one possible exception. A small fraction of diabetics are allergic to bovine or porcine insulin, the principal animal sources of the hormone. Human insulin, which differs from the animal hormone by a few amino acid residues produced by genetic engineering, has solved this problem.

Glucagon is unambiguously important as a chemical mediator. It was discovered because of its ability to raise blood sugar, not a unique property but shared with other hormones such as epinephrine (adrenaline). Current opinion considers the adipocytes (fat cells) as the major target cells. Here glucagon causes the breakdown of storage fat to glycerol and fatty acids.

Biochemists of my generation have been admonished, "Do not waste clean thoughts on impure enzymes." Should one add, "or contaminated reagents"? As this chapter illustrates, following such advice might have led to missed, or at least delayed, major discoveries.

Enzyme Suicide

An "accident" that took place in my laboratory, fortunately in retrospect, will serve as a final example of "the importance of being contaminated."

While studying the mechanism of unsaturated fatty acid synthesis in the

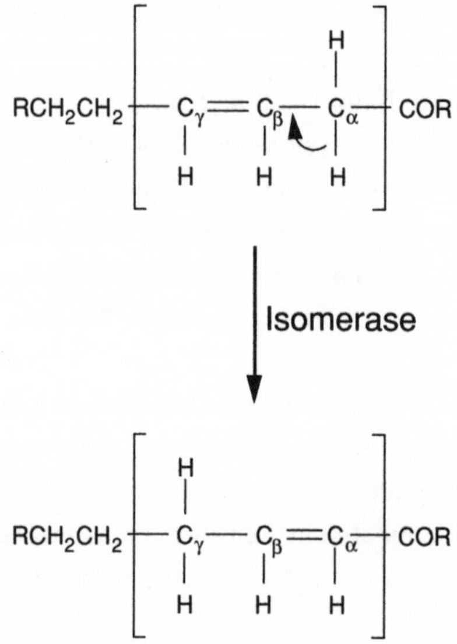

Fig. 4.5 In some bacteria the enzyme isomerase catalyzes a key step in the biosynthesis of unsaturated fatty acids: the shift of a hydrogen atom from the β,γ position (I) to the adjacent α,β position (II).

bacterium *E. coli* (in the 1960s), my collaborators isolated an enzyme that catalyzed the reaction described below (among others not pertinent here). For simplicity's sake, we will call the enzyme "isomerase." In some bacteria it catalyzes a key step in the biosynthesis of unsaturated fatty acids, the shift of a hydrogen atom from the β,γ position (I) to the adjacent α,β position (II) (Fig. 4.5). Such intramolecular shifts are known as isomerizations. The necessary substrate I, for assaying the enzyme, was commercially unavailable; it had to be homemade. The chemical procedure chosen was to reduce an acetylenic substrate analog Ia to the corresponding olefinic I. (Fig. 4.6).

In several runs this chemical reduction of Ia to I always afforded active substrate for the enzyme: but one batch, ostensibly made by the same procedure, proved to be totally inactive in the isomerase assay. Why? Fortunately, a small sample of "good" substrate was still around. When it was added to the standard enzyme assay system along with the inactive specimen, isomerization was completely suppressed. Clearly, the new

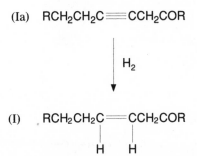

(Ia) $RCH_2CH_2C \equiv CCH_2COR$

H_2

(I) $RCH_2CH_2C = CCH_2COR$
                 | |
                 H   H

Fig. 4.6 Chemical synthesis of isomerase substrate.

substrate batch contained a contaminant, a very powerful inhibitor that prevented the left-over good substrate from reacting with the enzyme.

David Brock's detective work solved the puzzle. The above chemical procedure for acetylene reduction proved to be incomplete, yielding in this instance only 95 percent of the desired substrate (I). Some 5 percent of the starting material had escaped reduction of the triple bond and was the most likely culprit. When the pure acetylenic analog (Ia) was added to the enzyme, inhibition was rapid and complete after a few minutes.

George Helmkamp, then a graduate student, made a chance observation that was to clarify the mode of inhibition. He kept an alcoholic solution of the acetylenic compound Ia on his bench, routinely assaying it by an optical test (scanning its ultraviolet spectrum). To his dismay his alcoholic solutions of the inhibitor deteriorated gradually, judging by the optical test. Paradoxically, along with this spectral change, the acetylenic compound became an even more powerful inhibitor of the enzymatic isomerization. In all likelihood the acetylenic compound was not the inhibitor per se but a product derived from it, either spontaneously or on contact with the enzyme. All tests pointed to the latter. The enzyme promoted its own destruction and failed to distinguish between its normal olefinic substrate and the acetylenic analog, as outlined here (Fig. 4.7).

In both reactions the enzyme catalyzes the same chemical events, withdrawing a hydrogen (proton) atom from the α-carbon atom and adding a proton to the γ-carbon atom. In I, the result is a shift of the double bond from the β,γ to the α,β position, while in II the same process generates

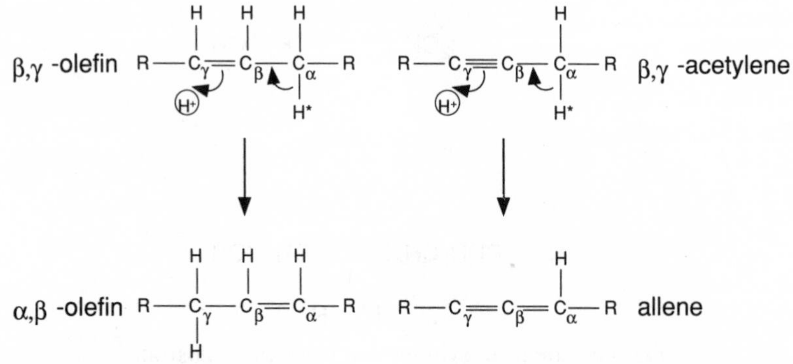

β,γ -olefin $R-C_\gamma=C_\beta-C_\alpha-R$ ... $R-C_\gamma\equiv C_\beta-C_\alpha-R$ β,γ -acetylene

α,β -olefin $R-C_\gamma-C_\beta=C_\alpha-R$ ... $R-C_\gamma=C_\beta=C_\alpha-R$ allene

Fig. 4.7 (I) Reaction of enzyme with normal olefinic substrate: olefinic isomerization. (II) Reaction of enzyme with acetylenic inhibitor: acetylene-allene isomerization.

a so-called allene, a structural element consisting of a carbon atom flanked by two double bonds. Crucial here is the failure of the enzyme to distinguish between the physiological olefinic substrate and the foreign acetylenic analog. The consequences of this lack of specificity are profound. Allenes are highly reactive chemicals and when facing a responsive enzyme will react irreversibly with the enzyme's active site—unlike the enzyme's physiological substrate.

We summarized these results with the following statement (Helmkamp et al., 1968): "Our experiments describe the rather unique case of an enzyme promoting its own destruction by catalyzing the transformation of a substrate analogue (the acetylenic compound) to an active site probe of extreme chemical reactivity (the allene)." While we let it go at that, other investigators of the same general phenomenon (Robert Abeles and Christopher Walsh) chose to call it enzyme suicide, popular terminology for a while. I myself have never adopted the term, for the following reason. Suicide is a deliberate act, "an instance of taking one's own life voluntarily and intentionally" (Webster, 2nd ed.) or "the horrid crime of destroying one's self by murder" (Samuel Johnson).

A more appropriate designation, free of any anthropomorphic connotation, might be a reference to a flaw of Nature, the inability of a biocatalyst to distinguish between friend and foe. The enzyme is victimized, deceived into binding, and transforms a foreign molecule with dire results.

These acts seal the enzyme's fate. Compounds in this category might more properly be called "Trojan horse inhibitors" or "wolves in sheep's clothing." Presently, "mechanism-based enzyme inhibitors" (or inhibition) has become the popular choice of enzymologists—not ideal, but a neutral and inoffensive term.

In our laboratory, Leon Kass tested the acetylenic enzyme inhibitor for its potential to combat bacterial infections, reasoning that certain biosynthetic pathways in bacteria are fundamentally different from those found in animals. Given the fact that unsaturated (olefinic) fatty acids are essential for the growth of E. coli and that the "isomerase" enzyme is responsible for their synthesis, one might expect that the acetylenic enzyme inhibitor would interfere with bacterial growth. This prediction was borne out (Kass, 1968). In quite low concentrations (10^{-5} M) the acetylenic analog retarded and eventually arrested the growth of E. coli cultures. Moreover, addition of the unsaturated oleic acid to the culture medium completely overcame the inhibition, (that is, restored bacterial growth), showing conclusively that the acetylenic analog's antibiotic effect was highly selective and not caused by general cell toxicity.

At that stage we entertained some cautious hopes that the acetylenic analog might be one of Paul Ehrlich's "magic bullets" (Zauberkugeln), targeting a single bacterial enzyme but nontoxic to animals because the animal body produces olefinic acids by totally different enzymes. The reasoning was correct, but Nature did not oblige. The animals' repertoire of enzymes is not at all the same as that of the bacterial world, a point made repeatedly throughout these essays. The acetylenic inhibitor, though bacteriostatic in E. coli cultures, failed to protect mice infected with E. coli. In animals a liver enzyme quickly breaks down the acetylenic inhibitor, a fate that is all too common with drugs that are either rationally designed or, as in the above case, proved—post factum—to have a rational basis. In more recent years, several dozen inhibitors based on the enzyme suicide paradigm have been designed and have contributed to an understanding of enzyme mechanisms. So far only one of these inhibitors, described below, has proved successful in chemotherapy.

In November 1990, the New York Times published an article entitled "New Drug to Treat Sleeping Sickness, African Trypanomiasis." It went on to say that the drug Ornidyl had been approved by the Food and Drug

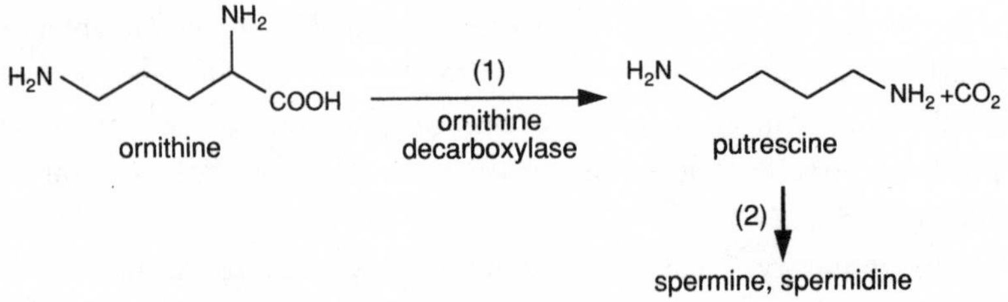

Fig. 4.8 Ornidyl inhibits the metabolism of the amino acid ornithine to the polyamines putrescine, spermine, and spermidine.

Administration (FDA) and the World Health Organization and manufactured after twelve years of research by the Marion–Merrel Dow Company.

Ornidyl, the original trade name for the compound difluoromethylornithine, $NH_2(CH_2)_2$—C—COOH, inhibits the metabolism of the amino acid ornithine (not a protein constituent, but a precursor of arginine) to the so-called polyamines putrescine, spermine, and spermidine, which in turn are essential regulators of DNA synthesis and hence cell growth (Fig. 4.8).

Reaction (1) is highly sensitive to the mechanism-based inhibitor Ornidyl, a "drug" operating on the general principle described above for the acetylenic analog. Remarkable, and critical to the chemotherapeutic efficacy of Ornidyl, the drug response varies substantially among species. The parasite's ornithine decarboxylase is highly sensitive to Ornidyl, whereas the same drug is well tolerated—without serious side effects—in human hosts. Differential drug response to infectious agents and the animal host is of course essential to successful drug design. The unprecedented success of penicillin in protecting the animal body against bacterial infections is also based on the different biochemical repertoire of animals and microbes. Penicillin interferes with the formation of bacterial cell walls, structures the animal body does not possess. Hence the essential absence of penicillin's side effects.

The nontoxicity of the antiprotozoan drug Ornidyl for humans has a fundamentally different basis. Ornithine decarboxylase, the enzyme's target for the drug, is functionally identical in man and parasite, yet one enzyme recognizes the drug and the other does not. In this instance the

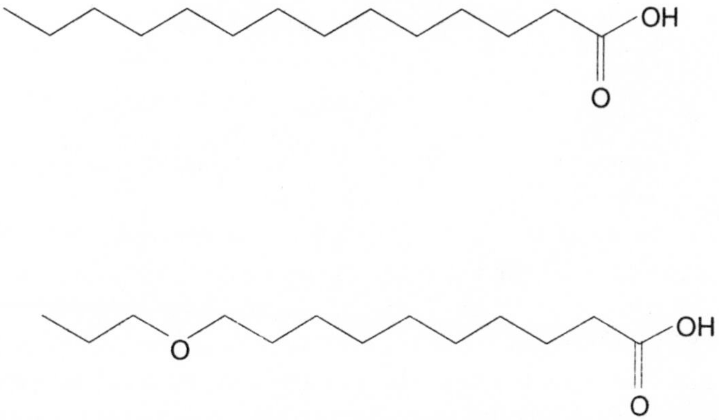

Fig. 4.9 A synthetic myristic acid analog, 10(propoxy)decanoic acid.

selective toxicity of the drug must be due to differences in the enzyme's structural details (so far unknown), which are not expressed at the catalytic level. For that reason, drug development in general must entail the testing of several unrelated animal species prior to clinical administration and FDA approval.

Sleeping sickness, caused by the protozoan *Trypanosoma brucei,* afflicts some fifty million people in the African continent, explaining the concerted search for selective drugs that affect the parasite but are nontoxic to the host. A promising alternative to Ornidyl is currently under development, again based on metabolic features probably not shared by host and parasite. Recently, trypanosomes have been shown to require, apparently uniquely, myristic acid, a fourteen-carbon fatty acid. Ordinarily, they obtain this molecule from the blood of the infected individual. The cells were killed rapidly when the normally required myristic acid was replaced by a synthetic myristic acid analog, 10(propoxy)decanoic acid (Doering et al., 1991) (Fig. 4.9). Once again, the analog is nontoxic to human cells. One would hope that similar chemotherapeutic approaches might eventually succeed in the treatment of other tropical diseases, notably malaria.

Frank Westheimer, in a 1984 address at the University of Maryland, aptly summarized the current progress in drug design: "Considering the great advances in therapy founded on empiricism, it is no small boast to claim that finally, drug design based on mechanistic enzymology can compete with man's vast storehouse of empirical information."

Bibliography

Oxysterols

1. A. Kandutsch, H. W. Chen, and H. J. Heiniger (1978), Biological activity of some oxygenated sterols, *Science* **201**, 498.

2. F. Taylor et al. (1984), Correlations between oxysterol binding to a cytosolic binding protein and potency in the repression of hydroxymethylglutaryl coenzyme A reductase, *J. Biol. Chem.* **259**, 12382.

3. M. S. Brown and J. Goldstein (1977), The low-density lipoprotein pathway and its relation to atherosclerosis, *Ann. Rev. Biochem.* **46**, 897; also in M. S. Brown and J. Goldstein (1985), *Les Prix Nobel* Nobel Foundation, pp. 158–198. Stockholm, Imprimerie Royale, P. A. Norstedt.

Contaminants of ATP; Vanadium

4. L. C. Cantley, L. G. Cantley, and L. Josephson (1978), Characterization of vanadate interactions with the (Na,K) ATPase, *J. Biol. Chem.* **253**, 7361.

5. R. L. Robson et al. (1986), The alternative nitrogenase of *Azotobacter chroococcum* is a vanadium enzyme, *Nature* **322**, 388.

6. R. Wever and K. Kustin (1990), Vanadium, a biologically relevant element, *Adv. Inorg. Chem.* **35**, 81–115.

7. L. A. Beaugé and I. M. Glynn (1977), A modifier of (Na^+, K^+) ATPase in commercial ATP, *Nature* **268**, 355–356.

8. L. C. Cantley, M. D. Resch, and G. Guidotti (1978), Vanadate inhibits the red cell (Na^+, K^+) ATPase from the cytoplasmic side, *Nature* **272**, 552–554.

9. I. G. Macara (1980), Vanadium—an element in search of a role, *Trends Biochem. Sci.* **5**, 92–94.

Phosphatidylcholine Synthesis

10. A. Kornberg and W. E. Pricer (1953), Enzymatic esterification of α-glycerophosphate by long-chain fatty acids, *J. Biol. Chem.* **204**, 345.

11. E. Kennedy (1962), The metabolism and functions of complex lipids, *The Harvey lectures*, pp. 143–171. New York, Academic Press.

Contaminants of Insulin

12. J. R. Murlin et al. (1923), Aqueous extracts of pancreas: Influence on the carbohydrate metabolism of depancreatized animals, *J. Biol. Chem.* **56**, 253–295.

13. J. Abel (1926), Crystalline insulin, *Proc. Nat. Acad. Sci.* (USA) **12**, 132.

14. D. A. Scott (1934), Crystalline insulin, *Biochem. J.* **28**, 1592–1602.

15. O. K. Behrens and W. W. Bromer (1958), Glucagon, *Glucagon Vit. Horm.* **16**, 263–301.

16. J. Audy and M. Keiley (1952), The content of glycogenolytic factor in the pancreas from different species, *Biochem. J.* **52**, 70.

17. E. W. Sutherland et al. (1949), Purification of the hyperglycemic-glycogenolytic factor from insulin and from gastric mucosa, *J. Biol. Chem.* **180**, 825.

18. C. Patzelt, H. Stager, R. J. Carroll, and D. F. Steiner (1979), Identification and processing of proglucagon in pancreatic islets, *Nature* **282**, 260–266.

Enzyme Suicide

19. G. Helmkamp, R. Rando, D. Brock, and K. Bloch (1968), β-hydroxydecanoylthioester dehydrase: Specificity of substates and acetylenic inhibitors, *J. Biol. Chem.* **243**, 3229.

20. K. Bloch (1986), The beginnings of "enzyme suicide," *J. Protein Chem.* **5**, 69.

21. L. Kass (1968), An antibacterial activity of 3-decynoyl-N-acetyl cysteamine, inhibition *in vivo* of β-hydroxydecanoyl thioester dehydrase, *J. Biol. Chem.* **243**, 3223.

22. A. Maycock and R. Abeles (1976), Suicide enzyme inactivators, *Accts. Chem. Res.* **9**, 313

23. C. Walsh (1977), *Horizons in biochemistry and biophysics,* vol. 3, p. 36. Reading, Massachusetts, Addison-Wesley.

24. R. Rando (1974), Chemistry and enzymology of k_{cat} inhibitors, *Science* **185**, 320.

25. A. Sjoerdsma (1981), Suicide enzyme inhibitors as potential drugs, *Clin. Pharm.* **30**, 3–22.

26. A. Petru et al. (1988), An analog of myristic acid with selective toxicity for African trypanosomiasis, *Am. J. Dis. Child.* **142**, 224.

27. T. L. Doering et al. (1991), African sleeping sickness in the United States: Successful treatment with eflornithine, *Science* **252**, 1851.

5

Receptors

Historical comments on the receptor concept are

presented including the predictions of a novelist about

the body's own opiates. Marijuana and digitalis glycosides

are discussed.

In this chapter, more than in others, I have disregarded the wise admonition "Cobbler, stick to your last." Never have I lectured on the subject of receptors, nor knowingly carried out receptor-related research. The reason for my transgression to unfamiliar territory may be contrived, but is perhaps justified by the unusual circumstances that directed my attention to the ligand-receptor phenomenon.

My interest in receptors came about inadvertently in 1987 while I was preparing a prefatory chapter entitled "Summing Up" for *Annual Reviews in Biochemistry.* I decided to refresh my memory of early days in my career, the period I spent at the Institute of High Altitude Research in Davos, Switzerland. This resort had two claims to fame. First, the ski runs from the Weissfluhjoch were the longest in the Alps, some as long as twenty miles, and second, Davos was the most popular European resort (*Kurort*) for patients suffering from tuberculosis. At the time, fresh mountain air was thought to be the only cure; antibiotics were not yet available.

In my high school Thomas Mann's *Magic Mountain* was required reading, and later I remembered that Davos was the locale of the novel. Mann
had spent long periods in Davos visiting his wife, a patient in one of the
sanatoria. The book was written in the period between 1915 and 1920, and
as I reread it very much later, in the early 1980s—the date is crucial—I
was startled to come across the following passage:

"Nothing different.—Oh, well, the stuff to-day was pure chemistry,"
Joachim unwillingly condescended to enlighten his cousin. It seemed there
was a sort of poisoning, an auto-infection of the organisms, so Dr. Krokowski
said; it was caused by the disintegration of a substance, of the nature of which
we were still ignorant, but which was present everywhere in the body; and
the products of this disintegration operated like an intoxicant upon the nerve-
centres of the spinal cord, with an effect similar to that of certain poisons,
such as morphia, or cocaine, when introduced in the usual way from outside.
"And so you get the hectic flush," said Hans Castorp. "But that's all worth
hearing. What doesn't the man know! He must have simply lapped it up. You
just wait, one of these days he will discover what that substance is that exists
everywhere in the body and sets free the soluble toxins that act like a narcotic
on the nervous system; then he will be able to fuddle us all more than ever.
Perhaps in the past they were able to do that very thing. When I listen to him,
I could almost think there is some truth in the old legends about love potions
and the like."

The passage struck me as extraordinary because Mann's Dr. Krokowski
had in fact predicted—one can say almost literally—the existence of the
intoxicating substances discovered sixty years later (in 1975), known today
as "the body's own opiates." Even more remarkably, the prophecy specified their origin "by the disintegration of a substance . . . and the products
of this disintegration operated like an intoxicant." Mann had in fact foreseen what today we call polyproteins, the precursors of endorphins and
enkephalins.

I shared my discovery of this passage with several neurobiologists, but
to my surprise none were aware of it, not even the discoverers of the
body's own opiates. I can only speculate why it had escaped their attention. Surely it was critical that I reread the novel in the 1980s, a few years
after the discovery of the opiate receptor (Pert and Snyder) and the

isolation of the enkephalins (Hughes and Kosterlitz). Fortunately, I was aware of these findings and had mentioned enkephalines to students during my last biochemistry lecture in 1980. Perhaps the more senior neuroscientists read "The Magic Mountain" in their youth, but probably not since. I must also conclude that those of the younger generation active in the field have probably never read the novel. Nonscientists would have failed to recognize the uniqueness of the prophecy.

Thomas Mann kept voluminous diaries all his life. Many of them he destroyed because they dealt with intimate personal details. But the diaries covering the period from 1915 to 1921, with their numerous references to *The Magic Mountain,* have been preserved. Nowhere do the diaries mention the sources that might have stimulated Mann to make his remarkable prophecy. Biology texts he cites as sources of information do not deal with the subject. Moreover, my correspondence with his son, Golo Mann, and the director of the Thomas Mann Archives in Zurich failed to provide any clues. The prediction of what we now know to be are an unusual class of metabolic signals and their mode of formation is unique in the annals of the life sciences,* and coming from a novelist, is even more remarkable.

In order to verify my "discovery," I wrote to Avram Goldstein of Stanford University, an expert on the subject of the body's own opiates. He called my attention to a contemporary paper by the French biologist M. Mavrojannis, entitled "Catalepsis of Morphine in Rats." In humans

---

*I recently learned of another equally remarkable prophesy. Bernard Witkop of the National Institutes of Health quoted the following prediction, made by the organic chemist Emil Fischer as early as 1914 on the occasion of the opening ceremonies for the first Kaiser Wilhelm Institute in Berlin: "We now are capable of obtaining numerous compounds that resemble, more or less, natural nucleic acids. How will they affect various living organisms? Will they be rejected or metabolized or will they participate in the construction of the cell nucleus? Only the experiment will give us the answer. I am bold enough to hope that, given the right conditions, the latter may happen and that artificial nucleic acids may be assimilated without degradation of the molecule. Such incorporation should lead to profound changes of the organism, resembling perhaps permanent changes or mutations as they have been observed before in nature" (Berichte **47** [1914], 3196).

hashish, a hemp (cannabis)-derived hallucinogen, has cataleptic effects, causing symptoms of nervous disorders such as prolonged muscular rigidity and the like. The noteworthy passage in Mavrojannis' paper reads as follows: "If one recalls [assumes?] that the organism fabricates normally narcotic substances, one must admit that in certain cases the cataleptic phenomenon is due either to an excessive production of narcotic substances or reduced elimination." Was Thomas Mann aware of the Frenchman's paper?

Technically, it was Mavrojannis who first conceived of the body's own narcotics. The publication date of his paper was 1923. The first edition of *The Magic Mountain* appeared in print in 1924. According to his diaries, however, Mann completed Chapter 5 of the novel, which contained his prophecy, on May 10, 1921. Unless he revised this chapter shortly before publication—and there is no hint that he did—the priority is clearly his. At any rate, one important detail of Mann's prescient passage remains unprecedented: the master molecule, of which the novel's Dr. Krokowski speaks, as a source of the endogenous intoxicating substances—the body's own opiates.

Apparently physiologists have long been intrigued by the puzzling phenomenon that plant substances such as the morphine contained in poppy seeds are pharmacologically active in the human body. Claude Bernard, one of the great nineteenth-century physicians, is said to have posed the problem, but I am unable to cite chapter and verse. Current opinion holds that plant alkaloids mimic, chemically as well as functionally, certain signaling substances that the animal body normally produces. Several examples have come to light. Active plant substances generally differ from those they mimic in one important respect. Morphine and the similarly acting cocaine are addictive, whereas the animals body's counterparts have a limited lifetime. They do not accumulate in amounts sufficient to cause the tolerance seen in drug addicts.

History of the Receptor Concept

In 1901 Paul Ehrlich published a drawing of cell surface membrane proteins (Fig. 5.1) that he later called "Receptors." It was his idea that

Fig. 5.1 Early diagram of cell surface membrane proteins drawn by Paul
Ehrlich. (From Ehrlich, *Schlussbetrachtungen: Erkrankungen des Blutes:
Nothnagels specielle Pathologie und Therapie,* vol. 3, Vienna, 1901.)

bioactive extracellular molecules were attracted to and bound specifically
to this entity.

Ehrlich's major claim to fame* rested on his early, successful attempt
to design "rational" drugs leading to the invention of the antisyphilytic,
salvarsan. But he was perhaps proudest of coining the term "receptor." It
was an outcome of his numerous researches on, inter alia, diphtheria toxin
and the parasitic trypanosomes. Ehrlich has been quoted (by Ernst Bäum-
ler) as saying, "After setting myself the task of penetrating into the life
process, I have succeeded in finding the key to it." It was a prophetic but
hardly a modest claim.

Ehrlich postulated several classes of receptors, distinguished by specific
binding sites for drugs, nutrients, toxic substances, and so on. Salvarsan,
an organic arsenic derivative and the first rationally designed and effective
drug** he referred to as a "magic bullet," aimed at and reaching its specific

*My sentimental interests in Paul Ehrlich are twofold. He was born in
Strehlen, a Silesian country town not far from my birthplace. Also, prior to
my emigration I found a haven, if briefly (1936), at the Paul Ehrlich Institute.
Privately endowed, it had not yet been shut down by the Nazi government.

**Salvarsan was originally named Arsphenamine, or Bayer 606, suggesting
that hundreds of compounds were tested during the search for an effective
and sufficiently nontoxic drug.

target. His motto, "Corpora non agunt nisi toccata"* (Bodies do not react unless they touch) guided his attempts to create what he intended to be a rationally designed drug. Today we speak of ligand-receptor interaction, a modern version of Ehrlich's maxim.

Significantly, Paul Ehrlich's contemporary and close friend was Emil Fischer, the renowned organic chemist and pioneer in biopolymer chemistry. To Fischer we owe the famed "lock and key" hypothesis, the earliest model for the mechanism of enzyme catalysis. Ehrlich adopted and extended Fischer's theory to explain the selective interactions between ligands and receptors. How he visualized them is shown in Fig. 5.1. By comparison, Fig. 5.2 illustrates a contemporary version typical for a wide variety of receptors. Common to both types of interactions is the specificity of the primary contacts between the substrate and the enzyme's binding site, and the ligand-receptor interaction. Here the analogy ends. Enzymes catalyze (that is, accelerate) the making and breaking of covalent bonds, bringing about chemical changes. By contrast, the events associated with the ligand-receptor interaction are physical, either translocation of the ligand from extracellular to intracellular space via a receptor, or else conformational, three-dimensional structure changes of the membranous receptor, resulting in transduction of the signaling message.

Modern Receptor Research

"The evolution of multicellular organisms (metazoans and metaphyta) has depended on the ability of cells to communicate with each other. Communication between cells is required to regulate their development and organization into tissues, to control their growth and division and to coordinate their diverse activities" (B. Alberts et al., *Molecular Biology of the Cell,* Garland Publishing, 1983, p. 717).

Among the numerous cellular signals, several classes can be distinguished:

1. One category comprises hormones—chemicals made by an endocrine

---

*According to Mason Hammond, Ehrlich's quotation is "linguistically mixed," a mélange of Latin and modern Italian.

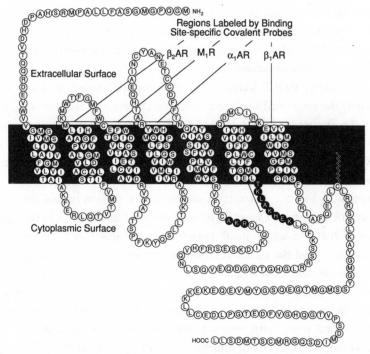

Fig. 5.2 Primary amino acid sequence of the human β₂-adrenergic receptor, indicating regions involved in ligand binding and G-protein coupling. The sequence of the protein is arranged about the darkened area (representing the plasma membrane) so as to depict the putative arrangement of the seven hydrophobic helical amino acid stretches crossing the plasma membrane. Both extracellular and cytoplasmic sides of the plasma membrane are indicated. (From R. F. Lefkowitz, *Membrane Proteins Targeting and Transduction,* Robert A. Welch Foundation Proceedings, Houston, Texas, 1989, p. 185.)

gland (one of the glands of internal secretion), such as the pancreas, pituitary, ovary, testes, and thyroid. These molecules, secreted into the bloodstream or lymph travel to distant receptors elsewhere in the body. Among the many examples are the steroid hormones, adrenaline, and thyroxin.

2. Other signaling molecules are known as local chemical mediators. They interact with receptors associated with cells in direct contact with

signal-producing cells, including for example histamine, which makes blood vessels leaky, and the prostaglandins, agents that cause smooth muscle cells to contract.

3. The neurotransmitters constitute a subcategory of locally acting signals, produced by and acting only on adjacent nerve cells. Among them are norepinephrine, differing from epinephrine (adrenaline) only by the absence of a methyl group,* and also the enkephalins, the pain-relieving "body's own opiates" acting on nerve terminals.

Known signaling molecules and their specific receptors probably number in the hundreds, and the end is not yet in sight. In spite of the enormous chemical diversity in size and structure of signals, there is a striking similarity in the overall structural design of the receptors (see Fig. 5.2). All receptors studied so far are large membrane-spanning glycoprotein molecules of which seven fragments or loops are embedded in the water-insoluble lipid phase of the plasma membrane, the barrier between external and internal cellular space.

The Body's Own Opiates

Widely used in receptor research are the terms "agonist" and "antagonist." Both agents bind to the same receptor in some but not all instances because of structural similarity. Agonists such as morphine activate the receptor productively for the ensuing pharmacological effects. Conversely, antagonists occupying and blocking the same receptor site interfere with subsequent signal transduction. A classic example is the discovery of the opiate receptor by Pert and Snyder (1973), accomplished with the aid of the morphine antagonist naloxone, a synthetic compound closely related to morphine but not occurring in Nature. It was chosen for technical reasons. Naloxone could be prepared in a highly radioactive form; the labeling of morphine presented much more of a challenge.

During the early 1970s, several research teams began a search for mol-

*This seemingly minor chemical change produces profound differences in function. Epinephrine receptors will not bind norepinephrine and vice versa, an example of exquisite receptor specificity.

ecules that would bind—that is, would serve as ligands for the opiate receptor. The laboratories of A. Terrenius in Stockholm, A. Goldstein at Stanford, and H. Kosterlitz in Aberdeen found such substances in the brain, which were designated "morphine agonists." My own belief is that the thoughts postulating the existence of the body's own opiates were in the air, and I find it difficult to assign priority to the ideas of one or the other of these investigators. To the best of my knowledge, awareness of Thomas Mann's prophecy more than fifty years earlier was not a guiding thought.

Techniques for finding "needles in a haystack" were well advanced in the 1970s, enabling these researchers to detect morphine agonists in the brain. Unlike morphine, the agonists were relatively small molecules, probably peptide in character. In a major advance, the laboratory of Kosterlitz in 1975 identified two pentapeptides, morphine agonists now known as Leu-enkephalin and Met-enkephalin. Shortly thereafter, Goldstein and colleagues (1979) added dynorphin, a thirteen amino acid peptide to the family of endogenous opiates. As it turned out, all three peptides were ultimately derived from a very much larger protein named pro-opiocortin, a polyprotein* that breaks down in the body to several functionally diverse hormones, including corticotropin (ACTH) and the melanocyte stimulating hormone β-MSH, in addition to the endogenous opiates. Thus Thomas Mann's prophecy of the body's intoxicating products of disintegration proved to have reality.

Opiate receptors are part of the body's normal machinery. They preexist whether or not the animal is exposed in advance to morphine or other "intoxicating" substances. A major mystery remains. Undoubtedly it was a reasonable notion that a drug, natural or man-made, mimics the chemical structure of an internally produced agent and hence enters a ligand-receptor relationship. Yet mimicry in the case of morphine and the body's own opiates, the enkephalins, appears to be only functional. Structural similarity is less obvious. Whatever chemical analogy exists between the plant

---

*The first polyprotein, discovered by Donald Steiner, proved to be a high-molecular precursor of the pancreatic hormone insulin.

opiates and the brain pentapeptides, it is probably confined to small regions of the respective ligands, as we shall see.

The search for this category of the body's own drugs will undoubtedly continue in spite of the lack of success in designing chemically modified enkephalins, hoped to be nonaddictive. While some are potent analgesics, the synthesized substances unfortunately are as addictive as morphine.

Is Morphine Itself Synthesized in the Animal Body?

Soon after the discovery of the enkephalins in 1975, neuroscientists raised the question whether morphine-like materials, truly related to plant alkaloid chemistry, might also be produced normally in brain tissue. Previously there had been no reason to suspect that the opiates morphine and codeine were made exclusively by higher plants. Surprisingly, there is now some evidence for a second class of the body's own opiates in addition to the enkephalins, namely morphine, codeine, and substances closely related and known precursors of these plant alkaloids. Five years before the enkephalins were discovered, Davis and Walsh (1970) published an intriguing hypothesis. Apparently their main goal was to explain a possible relationship between alcohol addiction and the equally addictive response to opium alkaloids. Their principal arguments were as follows.

Alcohol, metabolized by way of its oxidation product acetaldehyde, induces alterations in the metabolism of tyrosine-derived dopamine such that two molecules of dopamine react with formation of a proposed intermediate known as tetrahydropaparevoline (THP) or norlaudanosoline (Fig. 5.3). In turn, THP's tricyclic structure resembles that of morphine and is converted to morphine in the poppy plant. To buttress their hypothesis, Davis and Walsh incubated rat brain tissue with ^{14}C-dopamine, with and without acetaldehyde. A significant increase in the radioactivity of the morphine precursor THP was observed in the presence of acetaldehyde. The authors concluded that the conversion of THP to morphine-related alkaloids may also have occurred in their brain preparations.

Results obtained during the last five years in the laboratories of Goldstein at Stanford and of S. Spector at the Roche Institute have lent support

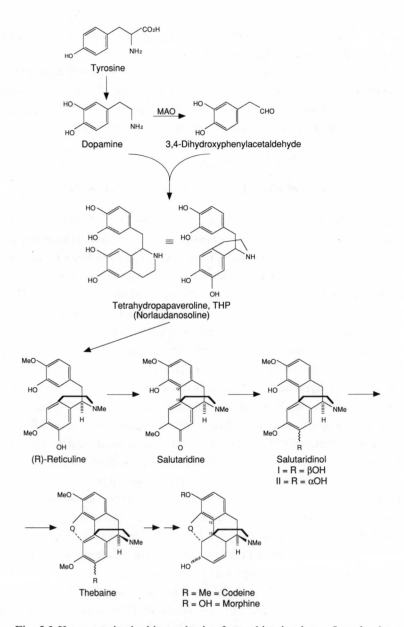

Fig. 5.3 Key steps in the biosynthesis of morphine in plants. In animals, only the formation of reticuline and later precursors of morphine have been demonstrated.

to the notion that some, if not necessarily all, steps of morphine biosynthesis* do occur in the animal brain. Precursors made in plants and closely related to morphine itself, such as salutaridine, thebaine, codeine, and the acetyl derivative of morphine itself, are present in mammalian tissue, albeit in trace amounts (Fig. 5.3). Moreover reticulene, a still-earlier morphine precursor, has been shown unequivocally to be converted to salutaridine in mammalian liver (Amann and Zenk, 1991). The experiments yielding these remarkable results were, and had to be, carefully controlled by excluding any dietary origin from plant matter of morphine or its precursors. The rigorous ruling out of such contaminations seems likely but needs confirmation.** Experience has shown that it is easier to prove the presence of a substance even in trace amounts than to prove its total absence. Bacterial sources (from the intestinal flora) have apparently not been considered.

At any rate, the phenomenon of the body's own opiates may have to be extended to the body's own morphine—literally, not just metaphorically. Apparently little attention has been given to the problem raised by this duality. Questions that remain, for example, include whether the quantities of the two types of endogenous opiates are reciprocally related—the same issue to be raised in Chapter 8 on alternative pathways. Second, do the enkephalins and the morphine-related molecules originate in the same or different regions of the brain? What are the control mechanisms responsible for the presumed production of one or the other of the two classes of endogenous analgesics? What are the stimuli?

As mentioned earlier, the original body's own opiates, endorphin and the enkaphalins, derive from a polyprotein proteolytically with the aid of cleavage enzymes named enkephalinases. How are these enzymes controlled?

Much remains to be learned about the response of opiate receptors, and

*To date, the conversion of ^{14}C-dopamine to morphine-like molecules has not been confirmed (A. Goldstein, private communication).

**Chemists will point out that the reactions producing morphine from reticuline, the earliest precursor found in animal tissues, are relatively simple: formation of and reduction of double bonds, epoxidation, and demethylation. These reactions may be catalyzed by nonspecific enzymes.

the level of endogenous opiates as a function of changing metabolic and environmental conditions. I have come across only one experiment along these lines, describing marked fluctuation of analgesic responses to noxic stimuli (Stevens and Pezalla, 1989). Hibernating squirrels were the test objects. In the summer the animals' reaction to noxious stimuli, such as placing acid on limbs, elicits an avoidance response, removal of the noxious agent by wiping. This response increases several-fold after the injection of morphine or the endogenous opiates. But in the fall, when hibernation commences, no such response occurs. The avoidance response is down regulated during hibernation, a state of torpor and metabolic inactivity.

There is a widespread popular notion that intense physical activity, jogging, or what is popularly known and advertised as "aerobic" exercise, creates a physiological high. Is it due to the release of endogenous opiates? Does the body respond similarly to acupuncture? The broader question of whether or not chemically defined phenomena underlie mind-body problems remains to be confirmed by adequately controlled experiments. If there is any substance to such perceptions, the question may be asked whether under various stresses the body produces enkephalins or, in fact, alkaloids of the morphine family.

Marijuana

Among the plant substances that alter human behavior, marijuana is one of the best known if not the most notorious. Its story began with isolation of the (hallucinogenic) principle from the dried flowers and leaves of hemp (*Cannabis sativa*), chemically characterized in 1964 by R. Mechaulam and named tetrahydrocannabinol (Δ^9 −THC). A long hiatus followed until the paradigm of the endogenous opiates stimulated similar efforts to search for a tetrahydrocannabinol receptor. By 1988 relatively specific receptors for Δ^9−THC had been detected in brain cells and cloned. The receptor resides in the cerebellum, hippocampus, and cerebral cortex. Mechaulam and colleagues have now found in porcine brain a putative endogenous ligand that binds to the marijuana receptor; they have named it anandamine (Fig. 5.4). The "body's own marijuana" will undoubtedly

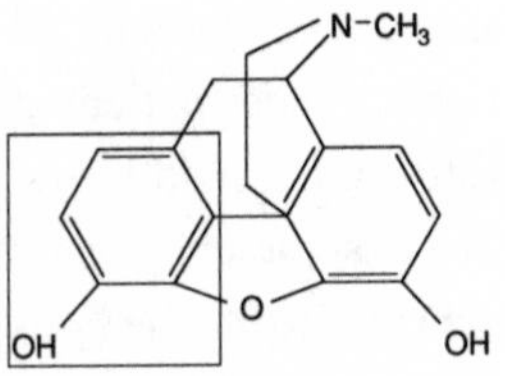

Morphine

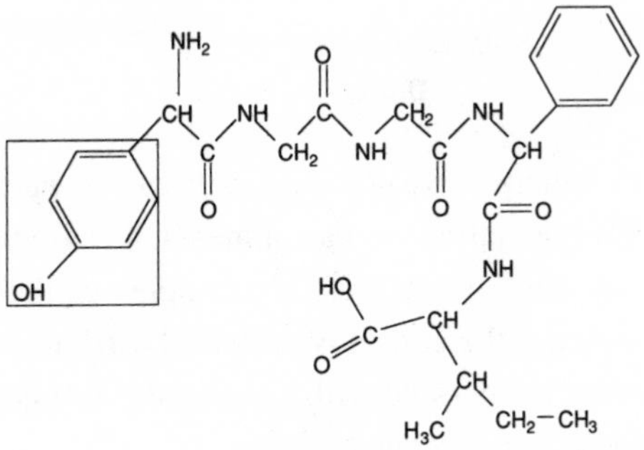

Leu-Enkephalin

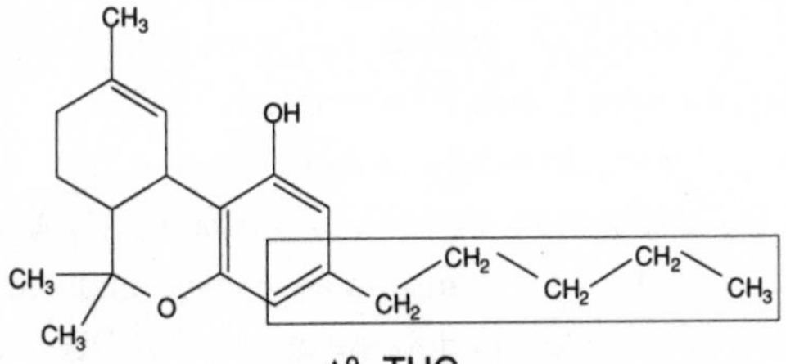

Δ^9 -THC

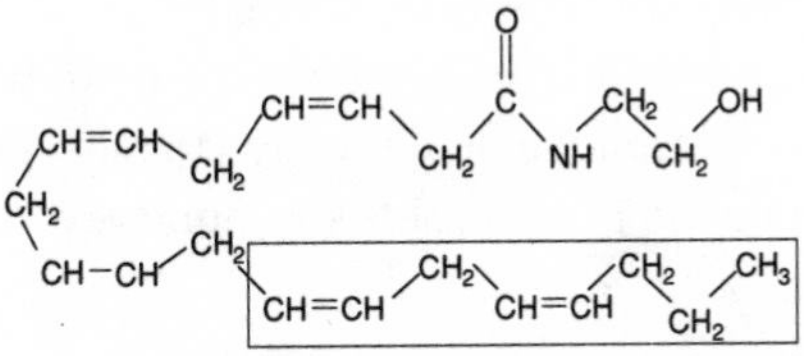

Anandamine

Fig. 5.4 Structural comparisons of the body's own opiates with morphine, and of the endogenous anandamine with the psychoactive molecule Δ^9-THC of marijuana. Only the boxed portions are identical chemically and therefore are likely binding sites for the respective receptors. (From Mechaulam et al., *Science* **258** [1992] 1882.)

attract much interest, given its beneficial effects such as relief from pain and vomiting, and stimulation of appetite in cancer patients, notably without causing addiction. As one investigator in the field notes: "Humans or other animals did not evolve a receptor for some chemical out of a plant; it is not just an accident that they fit [the receptor]. The body makes these receptors to accept chemicals that are important."

Digitalis

Since antiquity physicians have used extracts of the purple foxglove, *Digitalis purpurea,* for alleviating cardiac ailments, arrhythmia, and irregularities of heartbeat. The several active principles of digitalis, among them digitoxin and digoxin (the currently preferred cardiotonic drugs), are derivatives of cholesterol decorated with various sugar molecules and, as a class, known as digitalis glycosides.

During the past decade, digitalis glycosides were shown to bind with high affinity to certain cell membranes associated with sodium-potassium ATPases, the enzyme activities that regulate ion flux between extracellular and intracellular space. Digitalis inhibits this ion pump. Once again, following the "body's own opiate" precedent, a search began for endogenous inhibitors in the animal body that showed digitalis-like effects and qualified as natural ligands for the sodium-potassium ion pump. Not just one but several small molecules derived from human plasma had the anticipated properties, such as digoxin-like immunoreactivity (DLIR) and inhibition of the ion pump. Surprisingly, one of these structures proved to be an unusual phospholipid, composed of a long-chain, polyunsaturated fatty acid attached to a so-called lysophospholipid a molecule that lacks one of the two fatty acids phospholipids normally contain. In this instance, the ion pump inhibitor has no recognizable structural resemblance to members of the digitalis family.

However, a paper by Hamlin et al. (1991) describes the isolation from human plasma of the well-known ion pump inhibitor ouabain. This molecule is closely related to the foxglove-derived digitalis glycosides (Fig. 5.5). Plant ouabain (Fig. 5.5 A), which occurs in tropical strophanthus trees, binds with high affinity to the receptor for the digitalis glycosides and has powerful cardiotonic properties. The recovery of the digitalis-like

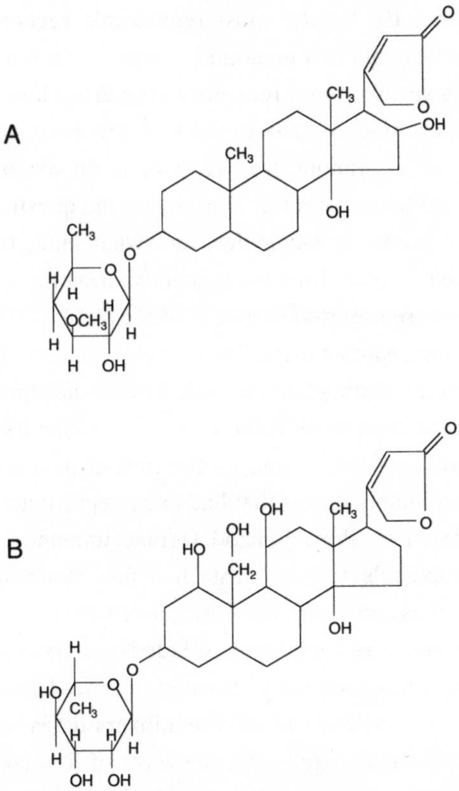

Fig. 5.5 Structural similarity of the plant digitalis glycoside (A) with ouabain (B), isolated from brain. (From Hamlin et al., 1991.)

ouabain (Fig. 5.5 B) from human plasma has been carefully controlled to show that it is of endogenous origin, not the result of dietary contamination. Undoubtedly, efforts will be initiated to isolate the enzymes responsible for the synthesis of ouabin in animal tissues. So far at least, the evidence seems convincing to regard ouabain as a prototype of the body's own digitalis.

The Body's Own Tranquilizers?

The previous discussion has centered on receptors for ligands of plant origin, the alkaloids of the poppy, digitalis from the foxglove, and hemp-derived marijuana. By far the most remarkable recent example of the receptor saga is the claim that man-made drugs, so far not found in Nature, can interact with specific brain receptors. The drugs known generically as diazepams or diazepines, currently one of the most widely prescribed class of sedatives or tranquilizers (valium is an example), bind to an apparently specific brain receptor. Once again the question arises whether the animal body produces molecules, and what kind, that show affinity for the diazepine receptor. Immunological experiments—which tend to be highly specific—were reported to give positive results (Mohler and Okada, 1977). The ensuing search for the "body's own valium" in rat and bovine brain identified a molecule identical with a benzodiazepine (N-desmethyldiazepam), a close cousin of valium. The investigators (Sangmeswaran et al., 1986) performed what was, on the face of it, a convincing experiment. Extracts of human brains that had been kept frozen and stored since 1940 were reported to show benzodiazepine immunoreactivity, that is, response to a molecule that chemists had first synthesized only in the 1960s and that was certainly unknown in 1940! Previous exposure of animals or humans was therefore excluded—or so it appeared. These astonishing results have not been repeated in other laboratories.

Historically, the development of tranquilizers dates to 1952, when the alkaloid reserpine was isolated from the roots of *Rauwolfia serpentina,* a shrub in India known for centuries to yield extracts capable of reducing hypertension. Reserpine may have served as a model for development of the modern tranquilizers librium and valium, but chemical similarity to reserpine is not obvious. Whether reserpine binds to the diazepine receptor has apparently not been tested.

Bibliography

Endogenous Opiates

1. K. Bloch (1987), Summing up, *Ann. Rev Biochem.* **56**, 1–19.

2. Th. Mann (1969), *The magic mountain, New York, Vintage Books,* Random House; originally published in 1924 as *Der Zauberberg,* Berlin, S. Fischer Verlag.

3. M. Mavrojannis (1923), Catalepsis of morphine in rats, *Compt. Rend. Soc. Biol.* **55**, 1092.

4. C. Pert and S. Snyder (1973), The opiate receptor: Demonstration in nervous tissues, *Science* **179**, 1011.

5. S. Snyder (1989), *Brainstorming, the science and politics of opiate research.* Cambridge, Massachusetts, Harvard University Press.

6. J. Hughes et al. (1975), Brain pentapeptides with opiate agonist activity, *Nature* **258**, 577.

7. H. Kosterlitz (1975), Has morphine a physiological function in the animal kingdom? *Nature* **317**, 671.

8. A. Goldstein et al. (1979), Structure of an extraordinarily potent opiopeptide, *Proc. Nat. Acad. Sci.* (USA) **76**, 6666.

9. C. Stevens and P. Pezalla (1989), Down regulation of endogenous opiates, *Brain Res.* **494**, 227.

Morphine Synthesis in Plants and Animals

10. G. Kirby (1967), Morphine biosynthesis in plants, *Science* **155**, 170.

11. V. Davis and M. Walsh (1970), Biochemical basis of alcohol addiction, *Science* **167**, 1005.

12. F. Donnerer et al. (1986), Presence of codeine and morphine in the rat, *Proc. Nat. Acad. Sci.* (USA) **83**, 4566.

13. H. Kosterlitz (1987), Biosynthesis of morphine in the animal kingdom, *Nature* **330**, 606.

14. H. Kodeira and S. Spector (1988), The convulsant opiate thebaine in mammalian brain, *Proc. Nat. Acad. Sci.* (USA) **86**, 716.

15. C. Weitz, K. F. Fault, and A. Goldstein (1987), Synthesis of the skeleton of morphine by mammalian liver, *Nature* **333**, 674.

16. T. Amann and M. Zenk (1991), Formation of morphine precursors, *Tetrahedron Lett.* **32**, 3675.

Marijuana

17. L. A. Matsuda et al. (1990), Structure of marijuana receptors, *Nature* **346**, 561.

18. W. A. Devane et al. (1992), Isolation and structure of the brain constituent that bonds to cannabinoid receptors, *Science* **258**, 1946.

Digitalis Glycosides

19. R. Kelly et al. (1985), Digitalis-like factors in human plasma, *J. Biol. Chem.* **260**, 11396.

20. A. Dasgupta et al. (1989), Novel digoxin-like substances from plasma, *Biochem. Biophys. Res. Comm.* **152**, 1435.

21. J. Hamlin et al. (1991), Ouabin-like compound from human plasma, *Proc. Nat. Acad. Sci.* (USA) **88**, 6259.

Benzodiazepine Receptors

22. H. Mohler and T. Okada (1977), Benzodiazepine receptor, *Science* **198**, 849.

23. L. Sangmeswaran et al. (1986), Purification of a benzodiazepine receptor from bovine brain, *Proc. Nat. Acad. Sci.* (USA) **83**, 9236.

6

The Chemical Structure and
Biological Function of Sugars

Nature has designed the simple sugar molecule trehalose

to protect cells or organelles against various

environmental stresses: heat, cold desiccation, and

osmotic stress. We explore why in insects the

blood sugar is trehalose, not glucose.

What is the essence of the structure-function problem, one that is central to biochemistry, and how can it be defined? We begin with the axiom or proposition that all molecular transformations in cells, the making and breaking of chemical bonds, are catalyzed by specific enzymes. A very few biochemical reactions also occur spontaneously, but several orders of magnitude less rapidly than those that are enzyme-catalyzed.

A catalytic act requires an encounter between the substrate (the molecule to be converted), and the enzyme (the usually very much larger protein catalyst). For a productive encounter or collision, the substrate must reach and bind to the enzyme active site so that the catalytic act, the transformation of substrate(s) to product(s), can take place. Rates of acceleration over the noncatalyzed reaction are measured by determining the relative quantities of substrate remaining and product formed. The rates vary from case to case, ranging from single numbers to factors of several million.

Apart from rate acceleration, the secret of enzymes is specificity. The catalyst will ordinarily transform only one of the thousands of different molecules that occur in the body, that is, only the one that fits perfectly into the enzyme's active site. Early in the twentieth century the renowned organic chemist Emil Fischer compared this specificity to the fit of lock (enzyme) and key (substrate), a concept or metaphor that has stood the test of time. In more modern parlance, the substrate docks at the active site.

In essence, the structure-function problem the experimenter faces is to define at the molecular level the participating structural features both of the substrate and of the recipient enzyme catalyst that transforms it. Once these features are identified, a chemical mechanism can be formulated. Since known substrate molecules as well as their specific enzymes number in the thousands and new ones are being discovered almost daily, mechanisms of enzyme action will ultimately be predictable. How long that will take is anybody's guess. When will we know, and who is to decide that "the mechanism of enzyme action" has been solved?*

Our discussion so far has emphasized the mutual structural relationship of enzymes and their specific substrates and *how* enzymes bring about rate accelerations. Both are mechanistic chemical questions.

A more esoteric, some would say Aristotelian, structural problem of interest to me over the years is best characterized as a "why" question. To take a simple example, can we rationalize why glucose and no other

---

*A biochemistry text by White, Handler, and Smith (4th ed., New York, John Wiley, 1968) notes the existence of an unusual prize established in 1796, to be awarded by the French Academy of Sciences "for the discovery of the mechanism by which a ferment [read enzyme] converts a fermentable substrate into products." At the suggestion of Jean-Marie Lehn, I wrote to the academy, inquiring whether the prize still exists. I learned that the award was changed in 1803 from one kilogram of gold to 3,000 francs. In 1804 the jury in charge considered none of the nominations worthy of the prize, and there are no records indicating that the award has been made since.

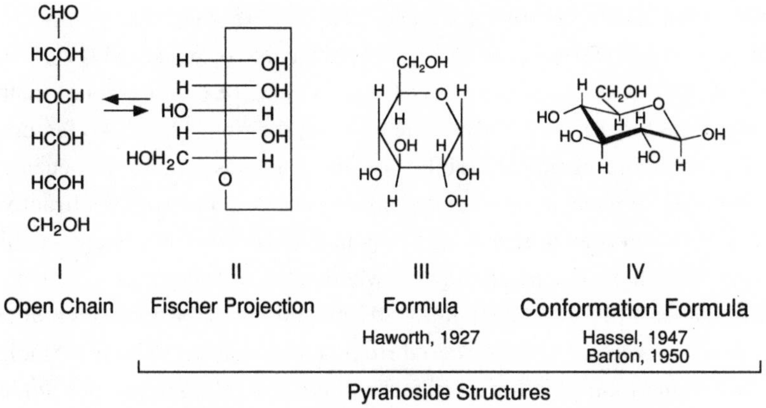

I	II	III	IV
Open Chain	Fischer Projection	Formula	Conformation Formula
		Haworth, 1927	Hassel, 1947 Barton, 1950

Pyranoside Structures

Fig. 6.1 Progressive refinement of the structure of glucose.

sugar has been "selected" as the universal carbon source for energy production, and to serve as the ultimate starting material for the construction
of all biomolecules? Was it accident, chemical necessity, or Nature's
wisdom? The six-carbon sugar can potentially exist in sixteen isomeric
forms, that is, sugars of the same molecular weight and elementary composition, differing only in the spatial arrangement of the H and OH substituents extending from the six-membered ring (Fig. 6.1). The question
"why" could not have been asked, let alone plausibly answered, until the
advent of conformational analysis (Hassell in 1947; Barton in 1950), approached first on theoretical grounds, then confirmed by x-ray crystallography.

Over the decades the written structure for glucose has been increasingly
refined. It was early recognized that in water solution the sugar exists in
two interconvertible forms, as an open-chain or an aldehydic molecule
(Fig. 6.1, I) and as a six-membered or pyranoside ring structure (II, III,
and IV), with the latter predominating.* Eventually, x-ray analysis provided a realistic conformational structure of crystalline glucose (IV) that
allows us to rationalize Nature's choice as follows.

Conformation means the disposition and orientation of atoms in three-

*The open-chain (I) and cylic (pyranoside II or III) structures interconvert
spontaneously by a shift of a hydrogen from C_5 to C_1.

dimensional space. In the six-membered glucose ring, the pyranoside structure (the preferred conformation) is chair-like or puckered (IV). Second, *all* the bulky substituents (OH groups) except one are equatorially oriented, that is, they lie in the same plane as the ring. These two features confer maximum chemical stability on the D-glucose molecule because, as chemists explain it, nonbonded interactions—repulsive or attractive forces between substituents—are minimized. Glucose is unique in this regard; all other six-carbon sugars, whether or not naturally occurring, contain at least one additional bulky OH substituent in the axial orientation (perpendicular to the six-membered ring) and are known to be less stable, or less prone than glucose to maintain their natural conformation under nonphysiological conditions such as heat and acidity.

Now we come to the key issue—highly speculative, to be sure. It must be assumed that prior to the beginnings of life organic molecules accumulated as a result of purely chemical (abiotic) processes. We do not know the environmental conditions that prevailed during the formation of organic from inorganic matter. Presumably they were very harsh by comparison with the environment prevailing after life began. At any rate, it is a plausible argument that, given very harsh conditions, the thermodynamically most stable compounds had the best chance to accumulate and survive. Glucose, the most stable of the sugars, was abundant and available for direct utilization by the earliest forms of life.*

Also, the vast majority of polysaccharides, the high-molecular-weight storage forms (fuel reserves) such as glycogen in animals and bacteria, and starch in plants, are polymers of glucose. Since specific enzymes are available for breaking down these storage polymers to glucose, additional steps are not needed for entry of the sugar molecule into the metabolic machinery.

In glucose four of the six carbons are asymmetric. Hence 2^4, or sixteen,

---

*Glucose—its unique properties and why it has been selected as the universal fuel and building stone for cells—is discussed in great detail by D. E. Green and R. F. Goldberger in *Molecular Insights into the Living Process* (New York, Academic Press, 1967). That description is well worth reading today. In my opinion, modern textbooks do not sufficiently emphasize the issue of why glucose was "selected" over its isomers.

isomers potentially exist. Yet apart from glucose only two of the isomers, mannose and galactose, occur in Nature. (A third, named altrose, has only recently been isolated from some microorganisms.) It strengthens the above argument that both mannose and galactose arise not de novo by pathways specific to these sugars, but by enzymatic isomerization of glucose, that is, replacement of axial or equatorial OH groups to an alternative orientation. This feature emphasizes the primacy and prominence of glucose in biological systems.

Metabolically, the much less abundant sugars such as galactose are specialized rather than central. They are not sources of energy as such, nor do they serve as building blocks or in polymeric form as fuel reserves. Instead the mannans, mannose polymers, serve structurally to strengthen the cell walls of yeast, fungi, and higher plants. Galactose polymers apparently do not exist. Still, galactose and mannose play an important—albeit specialized—role when attached to membrane proteins or lipids. Glycolipids and glycoproteins spatially orient the lipid or protein partners to which they are attached. All sugar molecules are hydrophilic (water seeking) and when linked to a membrane protein will position themselves in the aqueous phase surrounding the membrane.

Functionally important, the sugar residues of membrane-associated glycoproteins have "information content." Lubert Stryer (*Biochemistry,* 3rd ed., New York, W. H. Freeman, 1988) points out: "Carbohydrates have great potential for structural diversity. (1) An enormous number of patterns of sugar surfaces is possible because different monosaccharides (glucose, mannose, galactose) can be joined to each other through any of several hydroxyl groups; (2) the C_1 linkage can have either α or β configuration; (3) extensive branching is possible; and (4) many more different oligosaccharides (minipolymers) can be formed from a given number of sugars than oligopeptides from the same number of amino acids" (pp. 331–348). This structural diversity makes sugar residues residing on cell surfaces ideal for cell-cell recognition and interaction. Such interactions play an ever-increasing role in cell biology in general and more specifically in immunology, in the interaction of neural cells (for example, adhesion molecules) and in the binding of hormones to their specific glycoprotein receptors. Cell-cell interactions and glycoprotein function are developing, rather than mature, fields of research and still await definition in chemical

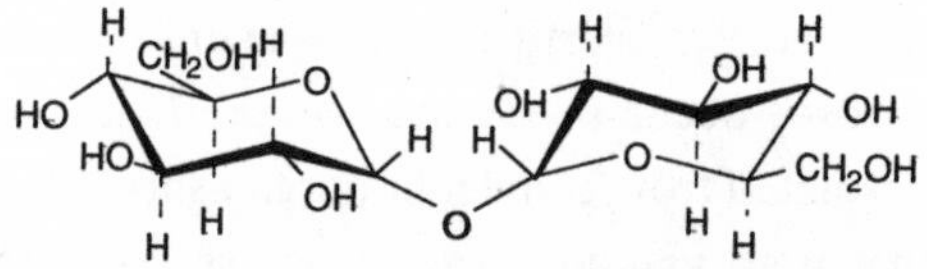

Fig. 6.2 The disaccharide trehalose contains two glucose units linked by an oxygen bridge between the two C_1 atoms in the α,α' configuration.

terms. Since sugar-based cell-cell interactions are reversible, they must involve the relatively weaker hydrogen bonds rather than the covalent bonds.

Another example of functional specialization involving sugars is the so-called antifreeze proteins occurring in the blood plasma of Antarctic fish. These glycoproteins of low molecular weight (10,000 to 20,000) contain disaccharides consisting of two galactose derivatives attached to the polypeptide backbone. Presumably they interfere with the nucleation of ice crystals.*

The purpose of this lengthy introduction to sugar chemistry, stressing the importance of conformational analysis, is to lay the basis for an appreciation of a remarkable example of structure-function relationships: Nature's selection of the disaccharide trehalose as a device for protecting cells and intracellular organelles against adverse environmental conditions.

Trehalose

Blood sugar commonly refers to glucose, the monosaccharide circulating in the bloodstream of all vertebrates. Not widely known, however, in most insects and some other invertebrates, the "blood sugar" is the unusual disaccharide trehalose containing two glucose units (Fig. 6.2; Wyatt and Kalf, 1957). Insect blood, known as hemolymph, is an open, freely circulating fluid in lieu of the vertebrate's vascular system for carrying

*A patent, borrowing from Antarctic fish physiology, proposes to transfer the genes for antifreeze proteins to citrus fruits, obviously for protecting frost-sensitive crops. I had the same idea—too late!

oxygen to the tissues. Hemoglobin, the familiar oxygen carrier, is generally absent in hemolymph. In some invertebrates (arthropods, mollusks) hemocyanin replaces hemoglobin.* Moreover, the various formed elements or corpuscles contained in vertebrate blood such as red cells (erythrocytes), white cells, lymphocytes, and monocytes (phagocytes) are absent in hemolymph or constitute at most a minor fraction of insect blood.

All these differences are probably irrelevant to the substitution of trehalose for glucose. Comparative biochemistry provides a more likely explanation. Though metabolized, trehalose is foreign to the vertebrate phylum. One finds this sugar in bacterial spores, for example of streptomycetes, fungi (mushrooms), and yeast, as well as in the hemolymph of many insect species, apparently with the exception of honey bees.** All unicellular organisms, as well as insects, notably at the larval and pupal stage, encounter and tolerate extreme environmental conditions such as desiccation (dehydration) and abnormally high or low temperatures. The ability to survive such environmental stresses correlates well with their content of trehalose. Experimental evidence that cells may owe these remarkable adaptations to trehalose has come both from applied an basic research.

Cell biologists facing the practical problem of preserving cells or cell fractions in a viable state for long periods do so by freeze-drying, also called lyophilization or cryoprotection—the lower the temperature, the better.*** Freezing and thawing of biological materials frequently diminishes

---

*Hemocyanin and hemoglobin are both oxygen carriers—that is, functionally identical; chemically, they are quite different and never occur together. Both are large proteins. Hemoglobin contains four atoms of iron, while the oxygen-binding metal in hemocyanin is copper. The oxygenated (oxygen-carrying) form of hemoglobin is dark red; the deoxy form, a red of lighter hue. Oxyhemocyanin is blue, and the oxygen-free form is colorless. When you squash a mosquito the red blood is yours, not the insect's.

**Honey bees, unlike most insects, practice a colonial rather than a solitary lifestyle. Bee colonies maintain a fairly constant temperature, about 65° F, insulating the cluster against major climatic changes. A fascinating account of the honey bee's lifestyle has appeared in the *New Yorker:* "Annals of Husbandry" by Sue Hubbell (May 9, 1988).

***Perhaps this idea occurred to a biologist who had learned that the now-

or even destroys biological activity, causing irreversible structural changes. For example, such treatments lead to fusion or collapse of vesicular membranes, resulting inter alia in the leakage of intracellular ions and metabolites.

Early on, when enzymes were first isolated in pure form and preserved by freeze-drying, some investigators observed, perhaps accidentally, a marked protective action of the polyhydroxy compounds glycerol and sucrose during freeze-drying and subsequent thawing. Years after this procedure had become a common practice, a variety of sugars including the disaccharides maltose, sucrose, and trehalose were tested and compared for their cryoprotective ability. Trehalose often proved superior to all other polyhydroxy compounds, especially at low concentrations. As a result, trehalose cryoprotection is currently used for stabilizing a wide variety of sensitive biological materials such as antibodies, vaccines, viruses, and embryos. The fact that trehalose is relatively resistant to alkali and acids is noteworthy but probably not relevant.

The distinctive presence of trehalose in lieu of glucose in insect hemolymph undoubtedly bears some relation to the inability of insects to control body temperature, in contrast to vertebrates.* If insects are to survive extreme temperatures or desiccation and escape the resultant irreversible damage to membranes or proteins, some agent must be present to compensate for the absence of liquid water. Lysozyme, an important enzyme protecting insects against foreign microorganisms, is easily heat inactivated; but adding trehalose to the enzyme raises the temperature at which irreversible damage occurs, by as much as 70° C. The suggestion has been made that trehalose mimics, or is in fact, "solid water," providing a number of hydroxyl groups that keep proteins and membranes in a pseudoaqueous environment. Many macromolecules actually tend to crystallize preferentially with one or more molecules of water. Proteins and DNA are examples.

---

extinct and probably prehistoric mammoth recovered from Siberian icefields was perfectly preserved and its meat unspoiled approximately thirty thousand years later.

*Insects, like reptiles, are not homeothermic. That is, they lack internal regulatory mechanisms for keeping their body temperature constant.

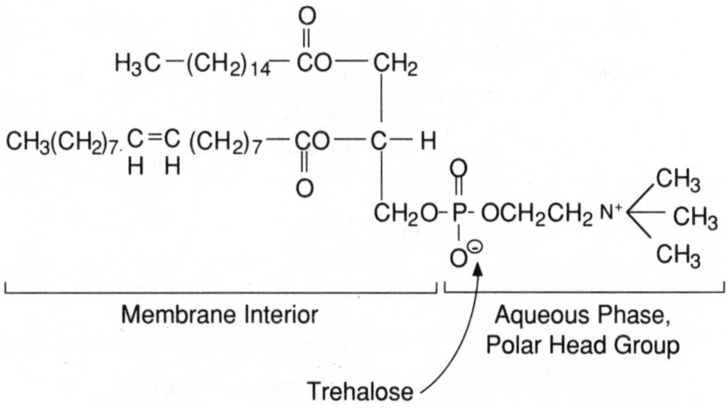

Fig. 6.3 Phospholipids contain long-chain fatty acids, glycerol, phosphate, and a nitrogenous (for example, choline in the text example).

A molecular explanation for the ability of sugars to prevent irreversible structural changes in macromolecules or membranes would probably invoke hydrogen bond formation. Water as well as sugars interact with the numerous hydrogen bond accepting groups provided by proteins and cell membrane components. Crowe and Crowe (1988), in studies with mixtures of sugars and phospholipid liposomes, tested the ability of cryoprotectant sugars to bind tightly, or "hold on," to water under conditions of extreme desiccation. But residual binding of water was not observed.

In the instance of trehalose-membrane interactions, which might conceivably extend from the sugar to the phosphate, the phospholipid's "polar head group" (Fig. 6.3) residual water bridging cannot be invoked. It would follow that sugar OH, "solid water," provides the likely hydrogen bond donors not only for membrane phospholipids but for proteins as well.

For some natural membranes, hydrogen bonding to phospholipids cannot alone explain the stabilizing effects of trehalose. Such is the case for thylakoids, the membrane-enclosed sites of energy conversion inside the chloroplasts. Unprotected isolated spinach thylakoid membranes rupture easily when frozen but remain intact when frozen in the presence of trehalose. Yet thylakoid membranes contain primarily galactosyl glycerides (glycolipids—combinations of sugars and fatty acids) and not phospholipids. It appears therefore that in thylakoids the protective or stabi-

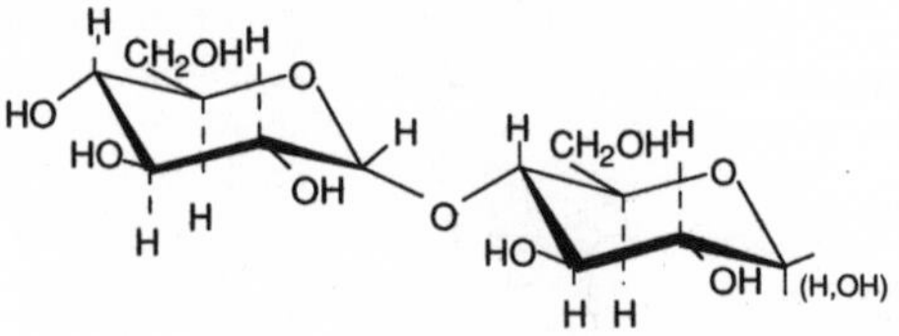

Maltose

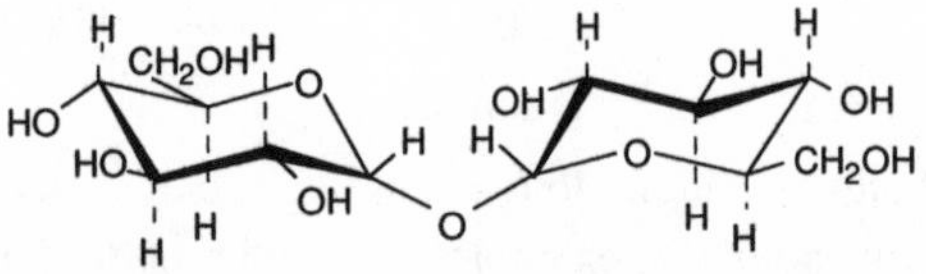

α,α'-Trehalose

Fig. 6.4 The disaccharide maltose (above) differs from trehalose (below) in the location of the oxygen bridges: 1,4 in maltose and 1,1 in trehalose.

lizing hydrogen bond interactions must occur between two sugars, trehalose and the sugar galactosyl residues of the glycolipids.

Why then is trehalose a highly effective cryoprotectant, and why have insects selected for their blood sugar this and not another disaccharide such as maltose? Maltose arising during the metabolic breakdown of glycogen or starch also contains two glucose units, in this case linked by a 1,4 glycosidic bond. In trehalose the corresponding bond links C_1 of one glucose unit to C_1 of the other (Fig. 6.4). The consequences of this structural difference are twofold. Probably the spatial dispositions of the hydrogen-donating OH groups in the two disaccharides, whatever they are, are not identical. Furthermore, the 1,4 glycosidic bond in maltose can open to generate spontaneously reducing and therefore reactive aldehyde groups at C_1. Ring opening of trehalose does not occur; therefore trehalose is nonreducing, or chemically inert.

As we have seen, glucose itself also exists partially in the open-chain or aldehyde form, which is reducing (Fig. 6.1). Therefore, in mammalian blood, glucose can react chemically with the terminal amino group of blood proteins, hemoglobin for example, to form a so-called Schiff base.

This reaction will block the amino terminal ends of the protein in an irreversible manner:

$$R_1CHO + NH_2 - R_2 \rightarrow R_1CH = N - R_2$$

R_1: Sugar moiety

R_2: Protein

Interestingly, Schiff-base formation between glucose and hemoglobin occurs in diabetes, when blood sugar concentrations are elevated. One of the essential roles of insulin is to keep blood glucose at low, nontoxic levels. This purpose may be achieved in part by sequestering excess glucose in the form of the Schiff-base adducts discussed above.* Insulin, produced in the pancreas of all vertebrates, is absent in insects.

While insulin-like molecules have been isolated from insect brain, their function in invertebrates is unknown, as is apparently insect diabetes. Trehalose, the insect's blood sugar, may reach levels of up to 1,600 mg percent in hemolymph, at least one order of magnitude greater than that of glucose in vertebrate blood, apparently without adverse effects. As noted, trehalose is nonreducing and therefore cannot undergo the Schiff-base reaction with proteins. During diapause, a nongrowing developmental insect stage perhaps comparable to that of microbial spores, the content of trehalose increases dramatically, providing protection against environmental stress (in this instance, desiccation). The choice of this disaccharide can therefore be rationalized, especially its selection over glucose and maltose, not only as a protective device against environmental stress but also because the molecule is chemically inert.

Trehalose, perhaps because it *is* chemically inert, most likely serves as the insect's energy store, the equivalent of glycogen in vertebrates or starch in higher plants. Thus, to meet flight muscle energy demand, hem-

---

*Several natural proteins contain a terminal N-acetyl group, instead of a free NH_2 group, a feature that has not been rationalized. Fetal hemoglobin (hemoglobin A_{1C}) is one such protein. Could this be a protective device for the fetus, preventing the glucose-hemoglobin interaction, which may be detrimental to hemoglobin's oxygen-binding capacity?

olymph trehalose (which, as noted, can reach very high levels) provides the major fuel initially, while during prolonged flight fat becomes the principal energy source.

Insects abide by the universal principle of using glucose as the carbon source for generating energy and as the ultimate molecule for building all cell constituents. This choice imposes a need for appropriate digestive enzymes. Insect trehalase serves this function, cleaving trehalose to glucose units for entry into the glycolytic pathway.*

Spores, the Dormant State

Microorganisms including bacteria, blue-green algae, yeasts, and fungi, when free-living or airborne (that is, when not infecting the animal body), tend to be exposed to a variety of stresses such as extreme temperature, dehydration, and changes in osmotic pressure. For their survival, Nature has designed a nonvegetative, nonreproducing form known as the sporal state. In the spore, or dormant form, which may last indefinitely, growth and development are temporarily suspended. Metabolic activity is minimal and the level of storage compounds high. Once the stress is relieved, spores return to the vegetative form. In order to remain viable, spores have adopted the same molecular strategy that allows developing insects to survive under adverse environmental conditions. For example, when *E. coli* is "shocked" by raising salt concentrations fivefold, which in effect partially dehydrates the bacteria (osmotic shock), the cells start synthesizing trehalose to counteract or prevent plasmolysis (cell disintegration).

Trehalose acts as an osmoprotectant in other microorganisms as well. In spores of streptomyces, bacteria producing the antibiotic streptomycin,

*Vertebrates can handle trehalose—though they have no recognized use for it—with the aid of similar hydrolytic enzymes found in the gut and in kidney tissue. If one gives credence to the Bible's story that the Israelites subsisted on manna in the wilderness, it would follow that they were able to digest trehalose and use it as an energy source. Mechanisms for the synthesis of trehalose appear to be absent in vertebrates.

the trehalose content rises up to fifty times that of the mycelial (vegetative) forms. Also, the heat stability at 60° C of the streptomycin spores increases by two orders of magnitude over that of vegetative cells. Spores of fungi, yeast, and soil-dwelling nematodes similarly respond to dehydration by synthesizing trehalose. When the osmotic stress is relieved by rehydration or return to normal temperatures and to rich media, spores return to the vegetative form. Germination and growth resume, along with reconversion of the trehalose stores to glucose.

Heat Shock Proteins

Various organisms respond to heating above their optimal growth temperatures by synthesizing a family of proteins known as heat shock proteins, or hsp's.

This adaptive response is under intensive study in bacterial cells, yeast, plants, and cultures of animal cells. Heat shock proteins are now recognized as a major adaptive device to counter environmental stress. As we have seen earlier in this chapter, insects, yeast, and bacterial spores respond to environmental stress, including heat, by initiating synthesis of the sugar trehalose. Are these two responses related in any way? The vast literature on heat shock proteins contains only a few reports dealing with apparent similarities between trehalose accumulation and the conditions producing heat shock proteins. Attfield (1987) first emphasized that "the pattern of heat-induced trehalose accumulation shows an interesting parallel to the heat shock response in yeast. Both responses occur rapidly, they are transient and decline when cells are returned to normal growth temperatures." Other parallels mentioned by Attfield include a requirement for RNA synthesis for expression of the two responses in bacterial spores.

Given the property of trehalose as a multifunctional protectant against heat, osmotic stress, and desiccation, one is tempted to pose the question whether stresses other than thermal also induce the synthesis of specific proteins analogous to heat shock proteins. Experiments along these lines would logically extend Attfield's insightful observations.

Structure and Function of Sugars

Discovery of Trehalose

In all likelihood, the chemist H. L. A. Wiggers was the first to lay hands on trehalose. In 1832 he published a long paper on the chemistry of "Mutterkorn" (ergot rye), the common fungal infestation of rye grass. This intriguing piece appeared in the first issue of Liebig'a *Annalen der Pharmazie*.* One of the fractions Wiggers extracted contained a "sweetish substance" and, as he put it, was an "eigenthümlicher Zucker" (peculiar sugar). Significantly, it was nonreducing, in contrast to sugars known at the time. If later workers attributed the discovery of trehalose to Wiggers, their evidence was rather weak, at least by today's standards; but probably they were correct in doing so. Ergot (*Claviceps purpurea*), as well as numerous fungi, proved later to be among the richest sources of trehalose, which explains its alternate but rarely used name "mycose" (*mykes,* Greek for fungus). Twenty years later (1852) the French chemist Berthelot, perhaps because he was interested in biblical manna, extracted a sweet substance from "trehala manna," contained in the cocoons of a lepidopterous Syrian beetle (family *Larinus nidificans*). This sugar he named trehalose. He found it nonreducing, like the sugar Wiggers had obtained earlier from ergot rye. Trehalose was apparently first characterized as the unusual α, α'-disaccharide in the 1920s. The advent of conformational analysis some thirty years later, together with x-ray crystallography, revealed the three-dimensional structure that should serve eventually as a basis for explaining its unusual properties.

The biblical manna (Arabic for "gift") said to have saved the Israelites from starvation during their journey through the Arabic wilderness** was

*The volume makes fascinating reading. It contains papers by Berzelius, Gay-Lussac, and several by Liebig himself. He was the sole editor of what apparently was his house organ. There are no indications whether any of the papers were refereed. Nearly all were by a single author. How times have changed!

**"And when the dew that lay was gone up, behold, upon the face of the wilderness there lay a small round thing, as small as the hoar frost on the ground. And when the children of Israel saw it, they said one to another, It is manna . . . And Moses said unto them, This is the bread which the Lord hath given you to eat . . . And the house of Israel called the name thereof probably the same sugar Berthelot isolated from beetle cocoons. But there are numerous possible sources for the sweet substance referred to as manna, apart from Berthelot's trehala manna. There is ash manna, Tamarisk manna (to this day collected by Sinai monks and erroneously sold as biblical manna), Persian manna, California manna, and perhaps others. The swett substance in the latter group contains instead of trehalose the sugar alcohol mannitol, also named dulcitol, derived from the monosaccharide mannose.

This chapter has discussed the diverse roles a relatively simple sugar plays in a variety of biological systems. Some explanations of why Nature chose it have been given. Yet the ultimate answer still eludes us. Even though the trehalose molecule has been inspected by state-of-the-art techniques including x-ray crystallography and nuclear magnetic resonance, we still cannot describe in molecular detail how the sugar protects cells from environmental stress.

One further point should be made. If biochemists' interests had been parochial, limited to studying human or mammalian physiology, trehalose would have been missed. From an anthropocentric point of view, trehalose is a vestigial molecule.

Bibliography

1. D. Wyatt and K. B. Kalf (1957), The chemistry of insect hemolymph: Trehalose and other carbohydrates, *J. Gen. Physiol.* **40**, 833.

2. G. Brown et al. (1972), The crystal structure of α,α' trehalose dihydrate from three independent x-ray determinations, *Acta Cryst.* **B28**, 3154.

3. C. Duda and E. Stevens (1990), Trehalose conformations in aqueous solution from optical rotation, *J. Am. Chem. Soc.* **112**, 4706.

4. P. Walian and B. Jap (1990), Trehalose stabilizes crystals of a porcine protein for low temperature ($-120°$ C) electron diffraction, *J. Mol. Biol.* **215**, 429.

Manna: and it was like coriander seed, white; and the taste of it was like wafers made with honey . . . And the children of Israel did eat manna forty years . . . until they came unto the borders of the land of Canaan" (Exodus, 16.

5. L. Crowe et al. (1987), Stabilization of dry phospholipid bilayers and proteins by sugars, *Biochem. J.* **224**, 1.

6. L. Crowe and J. Crowe (1988), Trehalose and dry dipalmitoyl-phosphatidylcholine revisited, *Biochem. Biophys. Acta* **946**, 183.

7. J. Crowe et al. (1987), Preservation of dry liposomes does not require retention of residual water, *Proc. Nat. Acad. Sci.* (USA) **84**, 1537.

8. D. Hincha (1989), Low concentrations of trehalose protect isolated thylakoids against mechanical freeze-thaw damage, *Biochem. Biophys. Acta* **987**, 231.

9. P. A. Attfield (1987), Trehalose accumulates in *Saccharomyces cerevisiae* during exposure to agents that induce heat shock responses, *FEBS Lett.* **225**, 259.

10. M. McBride and J. Ensign (1990), Regulation of trehalose metabolism by *Streptomyces griseus* spores, *J. Bact.* **172**, 3637.

11. M. J. Schlesinger (1990), Heat shock proteins, *J. Biol. Chem.* **265**, 12111.

12. H. L. A. Wiggers (1832), Untersuchungen über das Mutterkorn *Sekale Cornutum, Ann. Pharm.* **1**, 129–182.

7

Catabolism and Anabolism

We look at an earlier belief that biosynthetic processes

were catalyzed by the activities of the same enzymes that

bring about the breakdown of larger molecules. Chance

observations disclosed novel biochemical devices and

disposed of the reversibility hypothesis.

This largely historical chapter, thoroughly familiar to its author, might be unnecessarily detailed and technical. Yet this history, perhaps more than any other, documents an important point: much of biochemical knowledge was gained by trial and error, not by rational design.

The fate of chemical compounds in biological systems, a subject known as intermediary metabolism, is twofold. When cells degrade nutrients to smaller entities, we speak of catabolism, the collective term for events that generate thermal or chemical energy. By contrast, anabolism refers to the synthesis of the myriad body constituents, constructive processes that require energy. In the steady state (the adult organism), anabolism and catabolism, energetically uphill and downhill respectively, are in equilibrium.*

*Early in the nineteenth century the Swedish chemist Jacob Berzelius de-

From early in this century and through the following fifty years, much was learned about fermentation, the conversion of sugars to lactic acid and eventually to CO_2. Similarly, the mode of oxidative breakdown of long-chain fatty acids to two-carbon fragments became known in outline. Both processes yield chemical energy in the form of ATP.* During the same period all of the twenty amino acids contained in proteins were identified, as were some of their catabolic routes.

Biochemical experimentation at the time appears primitive by contemporary standards. In essence, metabolic studies consisted of feeding a test compound A to animals and then searching for increased amounts of a postulated conversion product B in the tissues, blood, or urine. This balance method, as it was called, in spite of its obvious limitations, yielded some remarkable insights into catabolism that were confirmed later by more sophisticated techniques. For example, such in vivo experiments led to the distinction between "essential" and "nonessential" nutritionally dispensable amino acids. Moreover, on the basis of such studies, amino acids could be classified as glycogenic (increased deposition of glycogen) or ketogenic (leading to the urinary excretion of the ketone bodies aceto-acetate and acetone in the diabetic organism). Rudolf Schoenheimer, the eminent biochemist who developed the methodology for tracing metabolic pathways with the aid of isotopes in the 1930s, jokingly but aptly compared the balance technique with the operation of a Coke dispenser: "A nickel (A) inserted into a Coke machine appears to produce (be converted to) a bottle of Coke (B)." The whole animal is indeed a black box—or it was until the current era, when noninvasive techniques such as NMR** enabled investigators to follow certain metabolic pathways in the living body.

---

fined organic chemistry as "the science describing the formation and transformation of organic molecules in the body," a definition today more appropriate for biochemistry than for organic chemistry.

*Adenosine triphosphate, or ATP, discovered in 1929, has been called the "universal currency of free energy in biological systems" (its structure is shown at the start of Chapter 3). It is generated by many catabolic reactions. Cleavage of either of the two P-O-P bonds of ATP provides the energy needed for energetically uphill reactions, both mechanical and chemical.

**The abbreviation NMR stands for nuclear magnetic resonance. Regrettably, the word "nuclear," in whatever context, is now frowned on so NMR has been largely replaced by MRI, Magnetic Resonance Imaging.

What remained almost entirely obscure until after World War II, for want of definitive experimental techniques, was the nature of the biosynthetic steps required to produce the conversion products, especially macromolecules. Of course, there were speculations about processes such as the biosynthesis of glycogen or starch, the polymerization of amino acids to proteins, the elongation of short-chain to long-chain fatty acids, and, somewhat later, the assembly of nucleic acids from their purine and pyrimidine precursors. However, the properties of some isolated enzymes deluded biochemists into believing that biosynthesis involved the same chemical reactions that were employed in the respective catabolic processes, but in the reverse direction. At the time there was no dissent from these widely held notions. Chance observations—not reasoning—were eventually responsible for the abandonment of what today are regarded as simplistic concepts.

That earlier biochemists accepted reversibility is understandable. In the test tube (in vitro), some of the enzymes catalyzing polymer to monomer conversions could indeed be forced into reverse: monomer $\rightleftharpoons$ polymer. In fact, under some artificial conditions the equilibrium constant in favor of synthesis, a measure of reversibility, proved to be greater than unity. It did not and perhaps could not occur to investigators that reversibility of enzyme reactions in the test tube was incompatible with physiological control. For example, in fat and carbohydrate metabolism, catabolic reactions occur in response to signals indicating energy needs and resulting in depletion of reserves, while the synthesis of polymers serves to convert excess fuel energy into storage products, a replenishing of reserves. The two processes cannot be regulated by the same signal unless "running idle" would serve a physiological purpose. Nature has solved this problem by designing novel chemical devices employed for biosynthesis and by segregating catabolic and anabolic processes in separate cell compartments. Three examples will be given.

Fatty Acid Breakdown and Synthesis

The animal body derives two-thirds of the total energy it requires from the oxidation of fatty acids, sixteen or eighteen carbon atoms in length, shortening the chains by two carbon atoms at a time. The process takes

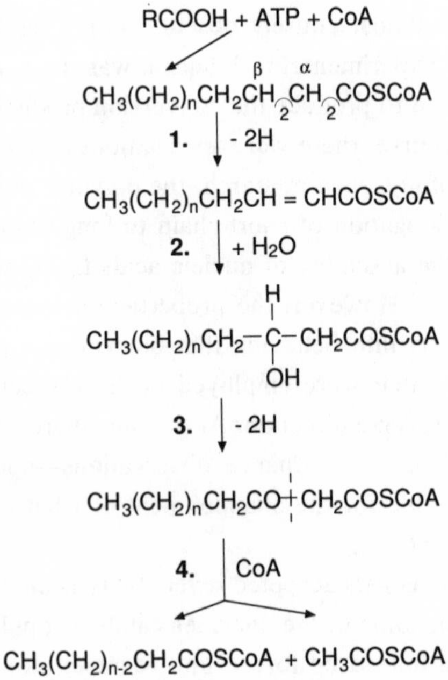

$$RCOOH + ATP + CoA$$

$$CH_3(CH_2)_nCH_2\overset{\beta}{C}H_2\overset{\alpha}{C}H_2COSCoA$$

1. $\downarrow$ - 2H

$$CH_3(CH_2)_nCH_2CH = CHCOSCoA$$

2. $\downarrow$ + H_2O

$$CH_3(CH_2)_nCH_2-\overset{\overset{H}{|}}{\underset{\underset{OH}{|}}{C}}-CH_2COSCoA$$

3. $\downarrow$ - 2H

$$CH_3(CH_2)_nCH_2CO \dashv CH_2COSCoA$$

4. | CoA

$$CH_3(CH_2)_{n-2}CH_2COSCoA + CH_3COSCoA$$

Fig. 7.1 Oxidative metabolism of long-chain fatty acids involves four enzymes for shortening the chains consecutively by two carbons. Palmitic acid (C_{16}) yields eight C_2 units iin the form of acetyl CoA.

place in the mitochondria, the organism's power plant, beginning with "activation" of the terminal carboxyl group of the fatty acid to a thioester: R-C-S-CoA (Fig. 7.1). Fritz Lipmann proposed this type of group activation in a seminal paper in 1941 (*Adv. Enzymol.* **1**, 99).

Four mitochondrial enzymes act sequentially to shorten a C_{16} or C_{18} ($n = 11$ or 13) fatty acid by the two-carbon fragment $CH_3COSCoA$ (acetyl CoA), repeating the four-step process until all of the sixteen or eighteen carbon atoms are dissembled to eight or nine two-carbon units.* How energy is derived from this process, known as sequential β-oxidation,

*The basic idea was developed and tested by F. Knoop (1904) long before the advent of isotopic tracers. He fed animals a series of fatty acids "labeled" by nonmetabolizable phenyl groups at the ω = carbon. The structures of the phenyl fatty acids excreted in the urine indicated that the fatty acid chains were shortened by two carbons at a time.

need not concern us here, except that the hydrogen atoms abstracted in reactions (1) and (3) of Fig. 7.1 fuel a combustion process generating many molecules of ATP. Further oxidation of acetyl CoA via the citric acid cycle, discovered by H. A. Krebs, furnishes additional ATP in the course of hydrogen combustion to water.

The biochemist Feodor Lynen summarized these results, first obtained in 1955, in what has become a biochemical classic (see Lynen, 1964). He was attracted by the idea that the same four mitochondrial enzymes might operate in reverse, producing as well as degrading long-chain fatty acids. In the biosynthetic direction if such a reaction occurred, hydrogen would be consumed, not produced, in reactions (1) and (3). In fact, by adding highly electronegative dyes as electron (hydrogen) donors, Lynen could reverse the process shown in Fig. 7.1 and obtain fatty acids up to ten carbon atoms in length.

Persuaded by this success, Lynen called his multienzyme system *fett-säure Cyclus*, assigning it the dual function of "cleaving the fatty acid chains to C_2 fragments, as well as the reverse to construct the long chains from C_2 units." In apparent support of his scheme he might have cited the earlier work of E. Stadtman and H. Barker (1949) with enzyme preparations from the microbe *Clostridium kluyvery*. This organism produces the much shorter butyric (C_4) and caproic (C_6) acids by the same general and reversible mechanism.

At the same time, it was obviously of concern to Lynen that one of the postulated steps, reaction (4) of Fig. 7.1, were it to proceed in the direction of bioynthesis, was highly unfavorable (*ungünstig*) energetically, with a cost of 8–9 kcal. At equilibrium the ratio of acetoacetyl CoA to acetyl CoA was as low as 1:250. Later, when Lynen had completed his masterly research on fatty acid synthesis in yeast (1961–1964), he conceded that he "was in error assuming that the apparent reversibility of the mitochondrial fatty acid cycle was responsible for fatty acid synthesis." The energy barrier of reaction (4) was too great. Still, Lynen's early concepts were universally accepted, taught, and published in all biochemistry texts until the early 1960s.

It is important that research addressing the same general subject be pursued simultaneously in more than one laboratory, and especially that young investigators not be deterred from doing so by dogma or the authority of more senior colleagues. Two such incidents are cases in point. Robert Langdon, a young postdoctoral fellow working in 1955 with a soluble fraction of rat liver—free of mitochondria—observed that for reversing steps (1) and (3) of Fig. 7.1 the reducing coenzyme was NADPH, not NADH as earlier assumed.* He concluded that while fatty acid oxidation may, as it does, proceed in the oxidizing atmosphere of the mitochondria, the cellular synthesis of fatty acids occurs predominantly in the reducing environment of the soluble cytoplasm. As mentioned, segregation of anabolic and catabolic pathways in different cellular compartments makes a great deal of sense physiologically.

Even more crucial to the ultimate chemical solution of the problem was another fortuitous observation. S. Wakil and colleagues (1959), working in David Green's Institute of Enzyme Research at the University of Wisconsin, were testing the effect of various buffers (milieus)—a common practice to optimize reaction rates. Bicarbonate (HCO_3^-) proved to be far superior to phosphate buffer.** As it turned out, CO_2 proved to be an absolute requirement in a liver system similar to that described by Langdon. Yet when radioactive $^{14}CO_2$ in the form of bicarbonate was added to the reaction mixture, not a trace of isotope appeared in the isolated fatty acids. Equally significant was Wakil's observation that one of the enzyme fractions active in fatty acid synthesis contained the vitamin biotin. By contrast, fatty acid oxidation did not require this vitamin.

No progress was made in the various interested laboratories for over a year. But discussions during a Gordon Conference in June 1958 set in motion a series of events that quickly led to resolution of the hitherto puzzling observations. At this conference David Green summarized the findings to date relevant to fatty acid synthesis. In the ensuing discussion— no printed record exists—Lynen, according to his account (1964), advanced the hypothesis that the "requirement for CO_2 (HCO_3^-) could be explained by the intermediary formation of malonyl CoA."

---

*The structures of the coenzymes will not be given here. Suffice it to say that NADPH and NADH, so-called pyridine nucleotides, serve as hydrogen donors or reducing agents, and NADP or NAD as hydrogen acceptors.

**This observation was confirmed in the laboratory of S. Gurin and also in my own. We did not pursue or publish it because we considered it trivial.

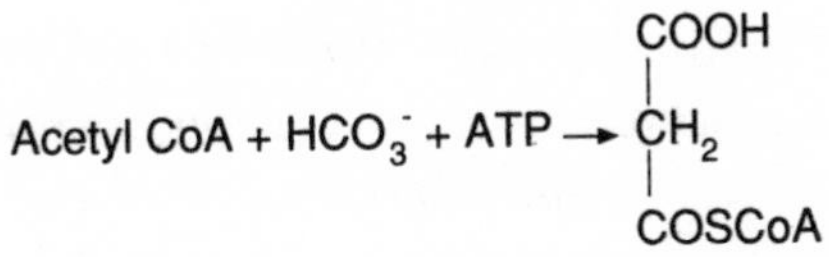

$$\text{Acetyl CoA} + HCO_3^- + ATP \rightarrow \begin{array}{c} COOH \\ | \\ CH_2 \\ | \\ COSCoA \end{array}$$

Malonyl CoA

Fig. 7.2 Formation of malonyl CoA provides the C_2 units for chain lengthening in fatty acid biosynthesis.

In short order, three papers provided experimental evidence for the hypothesis and explained why CO_2 is essential but not incorporated into the elongation product (Figs. 7.2 and 7.3).

R. Brady's report in August 1958 was the first to demonstrate the formation of malonyl CoA from acetyl CoA, HCO_3^-, and ATP. He also showed that this carboxylation product served as a precursor of long-chain fatty acids. Malonyl CoA, a doubly activated form of acetic acid, condenses readily with acetyl CoA, lengthening the carbon chain and rendering the process energetically favorable. The device serves to eliminate the energy barrier for the reversal of step (4) in Fig. 7.1. Shortly thereafter, in October 1958, Wakil took the problem one step further. He found that a biotin-containing enzyme fraction catalyzed the formation of "activated CO_2," now known as carboxybiotin. These results were verified by Lynen's laboratory in 1959.* Within half a year the basic chemistry of fatty acid synthesis had been elucidated. The process was shown to be unrelated mechanistically to fatty acid oxidation.

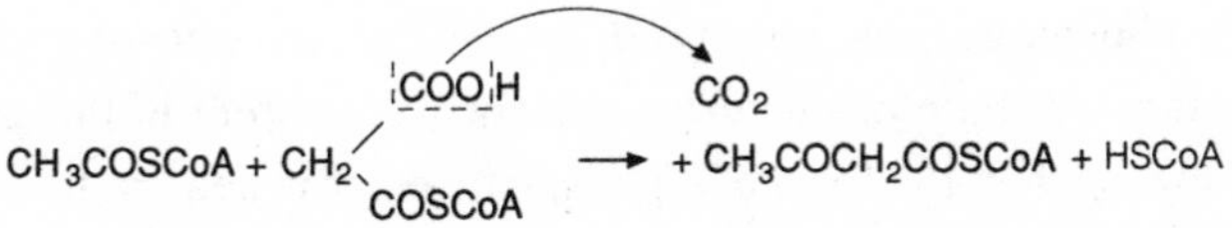

$$CH_3COSCoA + CH_2 \begin{array}{c} COOH \\ \diagdown \\ COSCoA \end{array} \xrightarrow{\quad CO_2 \quad} + CH_3COCH_2COSCoA + HSCoA$$

Fig. 7.3 The addition of C_2 units for elongating fatty acid chains.

*Malonyl CoA's existence as a natural product, but not yet invoked as a component in fatty acid biosynthesis, was known as early as 1955, when O. Hayaishi identified it as an intermediate in the decarboxylation of malonic acid by bacterial systems.

$$glycogen\ (n\text{-}glucose) + P_i \leftrightarrow glycogen(n\text{-}1) + glucose\text{-}1\text{-}phosphate$$

Fig. 7.4 Glycogen breakdown produces glucose phosphate by phosphorolysis.

Glycogen Synthesis and Glycogenolysis

Historically and intellectually, the subjects of glycogen breakdown (catabolism) and glycogen synthesis (anabolism) share many elements with fatty acid degradation and biosynthesis. Moreover, the resolution of both issues occurred during the same time frame. Both involve multiple, repetitive, and (it appeared at the time), seemingly identical steps chemically. The processes are reversible in principle, but not in physiologic reality. Eventually, as it turned out, in the biosynthetic process identical subunits are added repeatedly to a growing chain, malonyl CoA in fatty acid synthesis and an "activated" glucose molecule to saccharide starters.

Glycogen is a true polymer of large but variable molecular weight, whereas fatty acids are large (compared to other metabolites) but are not macromolecules. In contrast to nucleic acids and proteins, which are heteropolymers comprising twenty amino acids or four nitrogenous bases, glycogen and its plant analogs cellulose and starch are homopolymers made from single building stones.

The glycogen story begins with the fundamental discovery of Carl F. Cori and, independently, of Jacob Parnas in the 1930s that inorganic phosphate participates in the breakdown of glycogen. Cleavage of glycosidic bonds linking glucose units in the polymer is achieved by phophorolysis—not hydrolysis—leading to glucose-1-phosphate, or Cori ester (Fig. 7.4). This phosphorolysis, catalyzed by the enzyme phosphorylase, has the important attribute of capturing the energy inherent in the glycosidic bonds of glycogen in the form of a phosphate ester instead of free glucose. For this reason, little or no energy needs to be expended for further metabolic transformations.

Since glycogen is a macromolecule, the complete breakdown to glucose units requires numerous consecutive steps. It may be of help to the lay reader to know that the equilibrium constant of the reaction, a measure of the final concentrations of substrate and products (the ratio of substrate, glycogen, to product, glucose-1-phosphate, is inter alia a function of the ratio of inorganic phosphate to glucose-1-phosphate (or glucose-1-P). These ratios can be measured in the test tube with isolated phosphorylase enzyme. When the ratio is high, glycogen breakdown is favored; when it is very low, the reaction will more readily proceed from right to left, in the direction of glycogen synthesis. In vivo (in liver and muscle) the concentration of inorganic phosphate to glucose-1-P is ordinarily very large, favoring glycogen degradation. But in the test tube the concentrations of reactants can be manipulated—they are under the control of the experimenter. Thus when purified enzyme (phosphorylase) became available, the device of keeping the ratio of inorganic phosphate to glucose-1-P very low enabled investigators to force the reaction to proceed in the direction of glycogen synthesis.* No wonder that the prevailing views about the reversibility of glycogen breakdown went without challenge! No alternative mechanisms were proposed. Once again it took the uncommitted view of an investigator not explicitly working on glycogen synthesis to revolutionize the field.

Luis F. Leloir, a physician trained in Argentina and broadly familiar with intermediary metabolism, began his independent research in Buenos Aires working in a private laboratory endowed by the Fundación Campomar. The trail of Leloir's research leading to the discovery of "activated glucose" is best described in his own words: "My associate Dr. Caputto had done some research on the mammary gland and had the idea that glycogen was transformed into the milk sugar lactose. We then decided to study lactose breakdown by the yeast *Saccharomyces fragilis,* with the idea that this would give us information on the mechanism of lactose synthesis. In fact, it did give us information, but only after a long and tortuous path" (1970). As discussed later, in the context of milk intolerance, the disaccharide lactose is digested by the enzyme lactase to the monosaccharides glucose and galactose. To learn how galactose was metabolized, Leloir took a cue from the research of Hans Kosterlitz in Aberdeen indicating that galactose is converted first to galactose-1-phos-

*Glycogen's complex structure is tree-like, containing branches as well as straight chains of varying length. For simplification this issue is not addressed here. It is of little consequence to the main point of our discussion.

phate. Leloir isolated the responsible enzyme, galactokinase, that catalyzes the reaction:

$$galactose + ATP \rightarrow galactose\text{-}1\text{-}phosphate + ADP$$

Whether he attempted to show that galactose-1-phosphate was the activated form of the sugar for lactose synthesis the literature does not
record; but had he done so, the experiment would have failed. What
motivated Leloir to take the next step he does not say explicitly, except
to state, that his work in this field was the result of unplanned research;
it developed from studies on the conversion of galactose into glucose. One
of these "unplanned" experiments may have been to follow the conversion
of galactose-1-phosphate to glucose-1-phosphate in the yeast strain he had
found so useful in studying galactose metabolism. Leloir undertook to
investigate this isomerization of sugar phosphates in relatively crude extracts. His first step, a time-honored biochemical practice, was to dialyze
tissue extracts (that is, to enclose active fractions in a semipermeable
membrane, for removing small molecules from the nondialyzable high-
molecular-weight enzyme).* In dialyzed yeast preparations the transformation galactose-1-phosphate $\rightleftharpoons$ glucose-6-phosphate failed to occur. Obviously a small molecule had been separated and removed from the
enzyme protein(s) by dialysis; moreover, this component was thermostable and obviously not an enzyme.

Barely a year went by before Leloir and his associates had purified the
thermostable material sufficiently to elucidate its structure. The novel
molecule, named uridine diphosphate glucose (or UDP glucose, or
UDPG),** was formed from uridine triphosphate (UTP) as shown in
Fig. 7.5.

*Dialysis to separate large molecules from small is the classic device for
establishing whether an enzyme functions in cooperation with a coenzyme
that in nearly all cases is not a protein. Lipmann's discovery of coenzyme A
is another example of the technique. Today sophisticated by technically simple
and more rapid chromatographic procedures separate molecules according to
size.

**For whatever reason, Nature has chosen uridine and not adenine as the
nucleotide base for activating glucose in glycogen synthesis. For other related

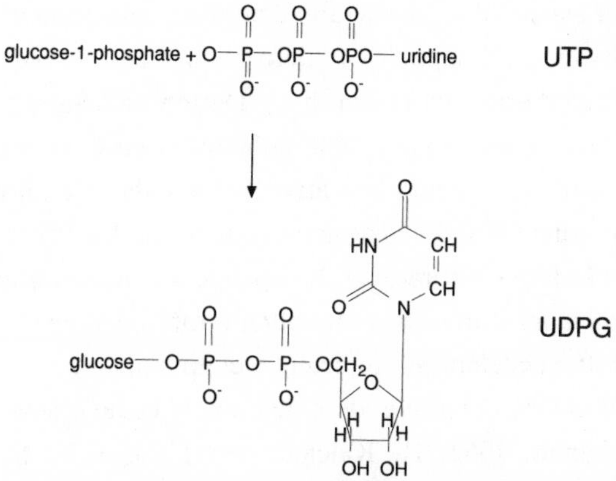

Fig. 7.5 The metabolically active form of glucose, UDPG,
and its precursor, UTP, in biosynthetic reactions.

Uridine diphosphate glucose proved to be the first sugar nucleotide or "activated" sugar, activated because it contains a pyrophosphate (or "high-energy") phosphate moiety. We have seen that the discovery of this co-factor for the galactose $\rightleftharpoons$ glucose isomerization was, in Leloir's words, "the result of unplanned research." Certainly, there was no inkling of its existence or preconceived notion of its structure. Similar findings, also serendipitous, and with different objectives, were made elsewhere. When Park and Johnson at the University of Wisconsin, microbiologists interested in the mode of action of antibiotics, added penicillin to bacterial cultures (*Staphylococcus aureus*), they observed "abnormal" amounts of an acid labile phosphate accumulating in the culture filtrate (1949). The new labile phosphorus compound contained one mole (molecule) of acid-labile phosphate, one mole of stable phosphate, 0.7 mole of glucose, and a moiety that had an ultraviolet absorption spectrum (at 262 mμ) consistent with the structure of a uridine component. This discovery, although preliminary, was the starting point for another monumental chapter, one that would deal with the biosynthesis of bacterial cell walls and the mode of action of penicillin (Park and Strominger, 1957).

---

polysaccharides, such as starch, which occur in plants and bacteria, the preferred glucosyl donor is adenine diphosphate glucose, ADPG. Why?

Still another aspect of sugar metabolism, the conjugation of glucuronate, a sugar carboxylic acid capable of "detoxifying" certain phenols, was the object of contemporaneous research by Dutton and Storey in Britain in 1951–1953. Once again, a dialyzable cofactor from liver was needed for forming the activated sugar derivative, which these authors suspected might have a structure somewhat analogous to the UDPG of Caputto and colleagues (in Leloir's laboratory). Independent contemporaneous discoveries in several laboratories are certainly not rare, tending to become more frequent with the acceleration of science communication.

Following Leloir's definitive identification of UDPG, several investigators (F. Buchanan, 1952; H. Kalckar, 1954) suggested that this sugar nucleotide may be the universal "activated" form in sugar transformations including the synthesis of disaccharides and sugar polymers. Still, it was Leloir's vision in addressing and undertaking all the critical experiments that established the specific role of UDPG in carbohydrate synthesis.*

Synthesis of Disaccharides and Polysaccharides

Reading between the lines, one can sense the ultimate aim and challenge of Leloir's research, the classic subject of glycogen synthesis. All other transformations of sugar nucleotides were in a sense preliminaries to this goal. While his studies of the galactose $\rightleftharpoons$ glucose isomerization reaction were the key to the discovery of UDPG, such hexose interconversions are freely reversible; that is, they require little or no input of energy in either direction.** The repetitive formation of glycosidic bonds is a different matter, as we have seen. Leloir approached it by first examining the disappearance of UDPG in various biological sources, including yeast,

---

*In earlier days key compounds or reactions were named for original discoverers (for instance, the Harden-Young ester and the Cori ester). Had the practice been continued, UDPG would undoubtedly and deservedly have been named the Leloir nucleotide. A very modest man, Leloir might well have rejected the honor.

**On the basis of free-energy changes alone, the need for high-energy intermediates in isomerizations is difficult to rationalize.

in the presence of putative glucosyl acceptors such as glucose-6-phosphate. This process was a kind of fishing expedition without any anticipation of the kind of fish that might be caught. The product turned out to be trehalose-6-phosphate, the phosphate derivative of the disaccharide that serves as the blood sugar of insects.* This disaccharide was the first example of a biosynthesis using UDPG and, notably, it was irreversible:

UDP-glucose + glucose-6-phosphate → trehalose-phosphate + UDP

In short order, an analogous example of disaccharide biosynthesis clarified by Leloir in 1953 was that of sucrose. Here UDPG serves as the glucosyl donor and fructose as the acceptor.**

Leloir was now ready to tackle his ultimate objective, the physiological mechanism of glycogen synthesis, even though it was widely accepted that the problem had been solved by the in vitro demonstration that glycogen breakdown is reversible. In 1959 Leloir detected a net synthesis of glycogen in rat liver preparations from UDPG and glycogen itself as an acceptor. He called the enzyme glycogen synthetase, distinct from phosphorylase.

$$\text{UDPG} + \text{acceptor}(n - 1) \rightarrow \text{acceptor}(n) + \text{UDP}$$

The reaction consists of one or more sequential additions of UDPG-derived glucose to a relatively large polysaccharide acceptor, thereby increasing its molecular weight.*** Most important, the reaction is irreversible (highly favored energetically), in contrast to that catalyzed by phosphorylase. Still it was necessary to ask whether Leloir's glycogen synthetase was solely responsible for glycogen synthesis under physiological

*See the structural formula of trehalose in Chapter 6.

**Sucrose, the most common disaccharide (from beets or sugarcane), happens to be the cheapest organic chemical, at a price of about ten cents per pound. It is pure coincidence that the chemical synthesis of sucrose was not achieved until the same year as its biosynthesis (R. Lemieux, 1953), in spite of decades of efforts by organic chemists. Moreover, the yield was disappointingly low, making chemical synthesis too costly for commercial production.

***The molecular weight of the polymer glycogen ranges widely. Since it is a storage material, its actual size or size distribution does not matter.

conditions. Two independent pathways to the same product are not uncommon in biochemical systems (see Chapter 8). There was no doubt that Carl and Gerty Cori had shown unequivocally that pure phosphorylase catalyzed glycogen formation, as well as breakdown, albeit under conditions that might be regarded as unphysiological.

Various observations cited by Leloir were indeed at odds with a role of phophorylase in glycogen synthesis in vivo. For example, Sutherland and Cori (1951) had earlier noted a glycogen breakdown to blood glucose in animals injected with the adrenal hormone epinephrine (adrenaline) as a result of *increased* phophorylase activity. If phosphorylase acted also as a glycogen synthetase, then adrenaline should stimulate this synthetic activity as well. This was not and obviously could not be the case.

Leloir's laboratory resolved the dilemma, showing that epinephrine *inhibited* his glycogen synthetase activity. The pancreatic hormone glucagon also produces hyperglycemia by stimulating phosphorylase, again only in the direction of glycogen breakdown. Clearly, these hormonal responses were not compatible with the notion that phosphorylase acts physiologically both as a glycogenolytic and as a major glycogenic enzyme.

Additional hormonal controls regulate blood sugar levels and glycogen storage. As everyone knows, the pancreatic hormone insulin lowers blood glucose, opposing the effects of epinephrine and glucagon. Later work has shown that one of insulin's effects is to stimulate Leloir's glycogen synthetase. Thus the same hormonal signals have antagonistic effects on phosphorylytic glycogen breakdown and on glycogen synthethis, a common molecular device for controlling, coordinating, and reinforcing anabolic and catabolic events.

If further proof for the separate roles of phosphorylase and glycogen synthetase was needed, it came from discoveries of several genetic disorders known as glycogen storage diseases or myopathies. In 1951 the physician B. McArdle described one of these, the disease named after him. He cited the case of an adult who suffered from extreme muscular weakness caused by impaired conversion of muscle glycogen to lactate, by way of glucose. The patient's blood sugar was far below normal and the content of muscle glycogen abnormally high. Later investigators (1959) showed that in such patients the muscle phosphorylase activity was less

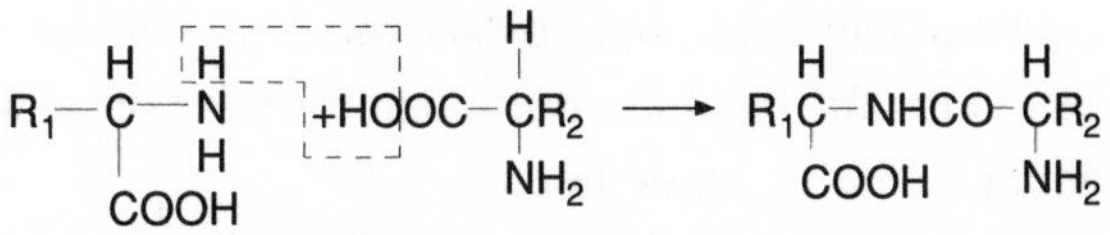

Fig. 7.6 The coupling of two amino acids to form peptides.

that 1 percent of normal. In other words, glycogen synthesis in afflicted individuals proceeds normally in the virtual absence of the enzyme responsible for glycogen breakdown.

It seems remarkable that this clinical evidence and the biochemical discovery of a separate glycogen synthetase surfaced—and reinforced each other—during the same decade.

Synthesis of Peptides and Proteins

For some years my laboratory investigated the enzymatic synthesis of glutathione, a tripeptide containing glutamic acid, cysteine, and glycine. We believed (naively, as it turned out) that glutathione might serve as a model for protein synthesis with respect to the mechanism for coupling the carboxyl group of one amino acid to the amino group of another. Few small-sized peptides were available or suitable for the purpose, and glutathione seemed the most appropriate because it contained one bona fide peptide bond (Fig. 7.6).

Concurrently, investigators addressed the same problem by studying the formation of amide bonds, for example in glutamine from glutamic acid and ammonia (John Speck) and the acetylation of sulfanilamide (Fritz Lipmann).* While peptide bonds are also amide bonds chemically, they refer more specifically to the linkage between the α-amino group of natural amino acids and the carboxyl group of its neighbors. All three lines of research with the potentially relevant model systems mentioned above led to the same result. Energy in the form of ATP is required—possibly

*Lipmann's studies in particular led—inadvertently—to the fundamental discovery of coenzyme A, which in turn proved to be the activating entity for carboxylic acids (except for amino acids in protein synthesis).

leading, as Lipmann had suggested, to "activated" or "energy-rich" amino acid carboxyl groups for reacting, presumably irreversibly, with an NH_2 group to form a peptide or amide bond.

Shortly after my arrival in Cambridge, Massachusetts, in 1954, I attended a cancer research conference held in Chatham on Cape Cod. The subject was protein biosynthesis. Paul Zamecnik, one of the principal speakers, summarized his recent research carried out largely with his collaborator Mahlon Hoagland. What we heard was to revolutionize the field of protein synthesis, and incidentally render irrelevant my own model experiments on small peptides and others on amide formation.*

Prior to the discoveries of Zamecnik's laboratory in the early 1950s, views on the mechanism of protein synthesis were in essence in line with the general belief that enzymes catalyzing lytic processes might be reversible. The same view was held, perhaps with some reservations, for protein synthesis. Thus, in the 1940s Max Bergmann, a widely recognized authority on peptide chemistry and the amino acid composition of proteins wrote as follows (in *Adv. Protein Chem.* **2** [1942], 49): "The proteins generated by cells under normal or pathogenic conditions must be the products of complex reactions in which many proteolytic enzymes and many substrates (the twenty amino acids) participate."** As an example, Bergmann cited the following process, known as transpeptidation, shown schematically in Fig. 7.7.

*A few years earlier I had participated in a similar conference, also dealing with protein synthesis, held at Pebble Beach Lodge on the Monterey Peninsula, California. Alfred Loomis, physicist, banker, and benefactor, sponsored and financed the meeting, which was attended by experts in the field. I roomed with Fritz Lipmann. At the conclusion Earl Evans, who had chaired the meeting, on behalf of the participants presented Loomis with the gift of a silver cigarette case. Thanking Loomis for his support, Evans noted that the outside of the case contained the engraved signatures of the participants; the inside—empty—held what was known about protein synthesis.

**Perhaps Bergmann used the verb "participate" advisedly. He did not state categorically that proteolytic enzymes alone catalyze the synthesis of proteins or of peptide bonds.

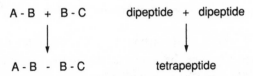

Fig. 7.7 Transportation, a rarely used mechanism for coupling amino acids to form peptides.

Reactions of this kind are indeed catalyzed; that is, they bring about chain lengthening in the presence of proteolytic enzymes such as chymotrypsin, papain, and intracellular protease enzymes known as cathepsins. Yet as Fruton and Simmonds point out (*General Biochemistry,* 1st ed., 1953, pp. 622–628), "Although these findings indicate the ability of proteinases to catalyze the elongation of peptide chains in a specific manner under physiological conditions, it must be emphasized that in the catalysis of replacement reactions, a protease [proteolytic enzyme] acts at *preformed* CO—NH bonds. If a biological system is provided only with free amino acids from which it must make proteins, some CONH bonds must be formed *de novo* in endergonic reactions, coupled with energy-yielding processes before the elongation of peptide chains by transpeptidation can occur." Thus transpeptidation alone could not account for de novo protein synthesis.

Transpeptidation has survived as a biologically significant event in at least one very special case: the biosynthesis of bacterial cell walls (Park and Strominger, 1957). One of the numerous reactions in the synthesis of this complex three-dimensional structure involves cross-linking of so-called peptidoglycan strands. One peptide bond is exchanged for another with little expenditure of energy, so it is a transpeptidation (as shown in Fig. 7.8), not a de novo formation of additional peptide bonds. The participating alanine is of the D or unnatural configuration not found in proteins.

glycine + D-alanyl-D-alanine ⟶ glycyl-D-alanine + D-alanine

Fig. 7.8 Transpeptidation in the biosynthesis of peptidoglycans, the cell walls of bacteria.

The New Era of Protein Synthesis

Soon after World War II carbon 14 (^{14}C), the relatively long lived radioactive carbon isotope, became available to the scientific community. Its discovery by Martin Kamen had been kept secret by the Atomic Energy Commission until the war ended, for whatever reason. The ^{14}C proved to be the ideal tool for studying biosynthetic reactions, either in intact animals or in cell preparations.

Paul Zamecnik entered the field of protein synthesis with an open mind, but as he later noted in his Harvey Lecture of 1959, "As a student of Bergmann at the [then] Rockefeller Institute, I felt a loyalty to his catheptic enzymes, but as a neighbor of Lipmann at the Massachusetts General Hospital, I developed a feeling that his concept of phosphorylated amino acid intermediates might be correct."

With the aid of the tracer technique, Zamecnik's laboratory soon demonstrated that ^{14}C-alanine and other labeled amino acids were readily incorporated into proteins of liver slices (1948). This incorporation required the presence of oxygen, besides which it was specific to natural amino acids; unnatural (D) amino acids failed to label the protein. To quote Zamecnik, "Both findings aroused our suspicion that the direct path from amino acid to protein was a new one, as yet undiscovered, not involving a reversal of protein synthesis." The oxygen requirement was later ascribed to the need for ATP generated by a process known as oxidative phosphorylation. Only with respect to the need of ATP for peptide bond formation did the model experiments, described earlier in this chapter, make a significant contribution.*

For studying biosynthetic processes in the test tube, the tissue slice technique was the only one then available and the first to be used in Zamecnik's laboratory. Its success depended on the investigator's manual skill in handling a razor blade. Besides, tissue slices had the major draw-

---

*In Chapter 3 I point out that all minimal processes essential for life, including protein synthesis, must have gone forward in the absence of oxygen. Only some specialized reactions were replaced by oxygen-requiring processes. Anaerobic forms of life generate ATP by fermentation, a much less efficient process than oxidative phosphorylation.

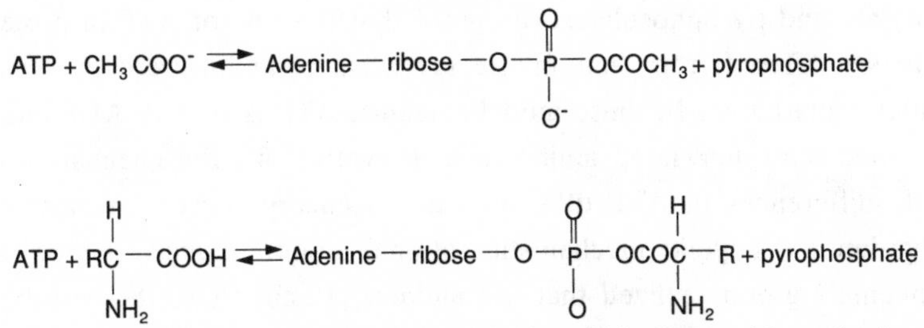

$$\text{ATP} + \text{CH}_3\text{COO}^- \rightleftharpoons \text{Adenine} - \text{ribose} - \text{O} - \overset{\overset{\text{O}}{\|}}{\underset{\underset{\text{O}^-}{|}}{\text{P}}} - \text{OCOCH}_3 + \text{pyrophosphate}$$

$$\text{ATP} + \overset{\overset{\text{H}}{|}}{\underset{\underset{\text{NH}_2}{|}}{\text{RC}}} - \text{COOH} \rightleftharpoons \text{Adenine} - \text{ribose} - \text{O} - \overset{\overset{\text{O}}{\|}}{\underset{\underset{\text{O}}{|}}{\text{P}}} - \text{OCOC} \overset{\overset{\text{H}}{|}}{\underset{\underset{\text{NH}_2}{|}}{}} - \text{R} + \text{pyrophosphate}$$

Fig. 7.9 Analogous reactions activate acetic acid in fatty acid synthesis and amino acids in protein synthesis.

back of excluding the entry of certain charged molecules, including ATP, into the intact cell. This permeability barrier precluded direct tests for the ATP requirement of biosynthetic reactions.

Nancy Bucher, a cell biologist and colleague of Zamecnik at the Massachusetts General Hospital, came to the rescue (1953). She developed a technique yielding biosynthetically active liver preparations by gently disrupting intact liver cells with the aid of a loosely fitting pestle (this detail was critical). Bucher first tested the procedure by incubating her homogenates with ^{14}C-labeled acetate, showing incorporation of this precursor into cholesterol.

For Zamecnik's pursuits, the Bucher technique had two major benefits. The participation of ATP in the formation of "activated" amino acids could be verified; in addition, cell fractions, or cytoplasmic fragments responsible for the subsequent steps in protein synthesis, could be separated and identified. Hoagland, taking advantage of the Bucher technique, showed that in an initial step the soluble fraction of the liver homogenate catalyzed the "activation" of various natural amino acids (L-isomers), forming phosphate derivatives in ATP-dependent reactions. D-isomers failed to react. Concurrent studies of Paul Berg, then at Western Reserve University, were crucial. He demonstrated that the carboxyl activation of acetic acid to acetyl CoA (a step preliminary also to the synthesis of fatty acids and cholesterol) proceeded by way of an acyl adenylate.

By analogy, as shown in Fig. 7.9, the carboxyl group of amino acids might be activated in a similar manner; and this proved to be the case (Hoagland, 1955). Also of importance, cleavage of ATP to amino acyl adenylate and pyrophosphate was proof that the role of ATP in protein synthesis differed from that in the synthesis of amides and the small peptide glutathione. In these model systems, ATP is split to ADP and a presumed phosphorylated amino acid derivative. To the chemist, such subtle differences in ATP utilization as an energy source for carboxyl activation are beyond mechanistic rationalization.* Equally important, Zamecnik's group realized that the aminoacyl anhydride, the carboxyl-activated amino acid, remains enzyme bound.

Perhaps the crucial step in protein synthesis distinguishing it from the synthesis of other macromolecules (glycogen, for instance) is that the product is not a polymer made from a single subunit (in other words, a homopolymer), but an ordered assembly of twenty individual amino acids differing from one protein to another in their relative proportions and the sequence in which they are arranged in the polypeptide chain. What predetermines that sequence and how is that sequence imposed on the condensation process? This question was asked during the middle 1950s by numerous investigators. Zamecnik focused on the ribonucleoprotein particle (ribosomes, containing RNA and proteins) as the earliest architectural site for completing the long peptide chains. In fact, the notion that nucleic acids take part in protein synthesis occurred to Caspersson in Sweden and Brachet in Belgium as early as 1940. It remained a hypothesis for more than a decade.

Beginning in 1955, efforts to test the role of RNA in protein synthesis were under way nearly simultaneously in half a dozen different laboratories. Zamecnik and collaborators provided perhaps the clearest and most significant results. Both ^{14}C-ATP and ^{14}C-leucine were shown to bind covalently to an unusually small cytoplasmic RNA named soluble RNA or s-RNA, now known as transfer RNA or t-RNA. The carboxyl-activated amino acid or aminoacyl adenylate had been transferred to a terminal locus on the RNA, entering unexpectedly into an ester linkage with a

---

*One can argue that the further cleavage of pyrophosphate into two monophosphates catalyzed by ubiquitous enzymes will pull the reactions shown in Fig. 7.9 in the forward (left-to-right) direction, thereby making them more efficient.

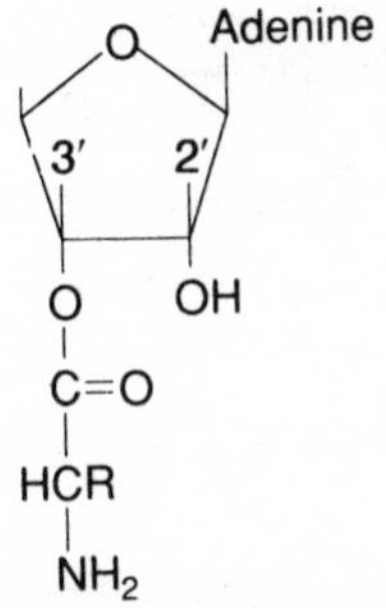

Fig. 7.10 Aminoacyl adenylate contains an ester linkage between amino acids and the 2'-OH groups in ribose.

hydroxyl group of the ribose (pentose) sugar of s-RNA (Fig. 7.10), catalyzed by the enzyme aminoacyl t-RNA synthetase.

Why was this mechanistic detail surprising, unforeseen by anyone? First of all, for decades biochemists had taken it for granted that peptides were formed by direct interaction between the free amino groups of one amino acid and the carboxyl group of another. If for energetic reasons one of the reactants needed to be activated, the energy-rich carboxyl phosphate $R-COPO_3$ or thioesters RC-SR were favored because precedents existed. At the time they were the only known examples of activated carboxyl groups. Structurally simple oxygen esters such as ethyl acetate were not regarded as energy rich, comparable to the pyrophosphate moiety (POPO) of ATP. However, as William Jencks (1969) points out, the free energy of hydrolysis of ester bonds (ester reactivity) depends critically on neighboring group effects, that is, of substituents adjacent to the ribose 3'-OH group to which the carboxyl component is attached. In aminoacyl t-RNA it is the neighboring 2'-OH group of ribose (Fig. 7.11) that is responsible for a fifteenfold to thirtyfold rate enhancement of reactivity over esters which lack this structural feature.

This argument gained support from experiments with model compounds (Zachau and Karau, 1960). It remains to be tested with the corresponding deoxyribose derivatives that contain H and C_2 instead of OH as neighbors to the O-ester group. One might argue that Nature's choice for creating a "high-energy" aminoacyl derivative fell on RNA (ribonucleic acid) and not DNA (deoxyribose nucleic acid). Such thoughts are certainly in line with

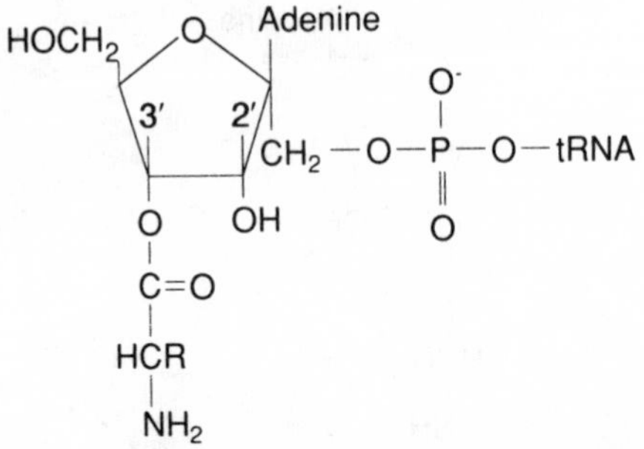

Fig. 7.11 Aminoacyl t-RNA, the ultimate amino acid derivative, entering peptides in protein synthesis.

increasingly popular hypotheses that RNA, and hence ribose, appeared earlier in evolution than the deoxyribose-containing DNA (Gilbert, 1986). In ATP, essential for protein synthesis, the sugar component is ribose, not deoxyribose. Furthermore, the only known pathway for deoxyribose is by conversion of ribose, further suggesting the evolutionary primacy of the latter.

Before we conclude this discussion of protein synthesis, one further pivotal contribution of Zamecnik's laboratory deserves mention. It began with the observation (1959) that "a single amino acid, e.g. valine, could be added to liver homogenates in increasing concentrations until a saturation level of labeling was reached." But the addition of ^{14}C amino acids other than valine resulted in further labeling of proteins. The two amino acids did not compete for attachment sites, leading to the important conclusion that there must be separate t-RNA molecules coding specifically for each of the twenty individual amino acids. In confirmation of this postulate, transfer RNAs, a family of twenty enzymes, have been isolated, each specific for acceptance of a single amino acid.

Formation of aminoacyl transfer RNA completes the set of reactions performed by the soluble portion of the cytoplasm. From here on, the particulate ribosomes take over. Ribosomes are particles, or organelles, containing ribonucleic acids of various kinds, including the informational "messenger" RNA formulated in 1961 by F. Jacob and J. Monod and at the same time by S. Brenner, F. Jacob, and M. Meselson. Moreover, ribosomes contain some fifty different proteins, elements that interact to

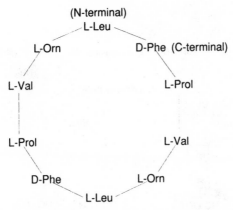

Fig. 7.12 Cyclic structure of the bacterial antibiotic gramicidin.

bring about the key chemical events in the terminal stages of protein synthesis. These include formation of peptide bonds and direction of the coded sequence in which the various amino acids are aligned along the polynucleotide chain. No attempt will be made to summarize such exceedingly complex processes. Voluminous treatises have dealt with these early achievements of molecular biology, epitomized by the discovery in 1953 of the double helical structure of DNA by Watson and Crick and the deciphering of the genetic code by M. Nirenberg and S. Ochoa in 1964.

Peptide Antibiotics

The discovery and structure determination of penicillin set in motion a concerted search for antibiotics both in academia and in industry. Some antibiotics proved to be cyclic peptides, useful for topical application, and others were insecticides. Of special interest academically was gramicidin S, a decapeptide produced by a strain of *Bacillus brevis,* containing both natural (L) and unnatural (D) amino acids* not found in proteins. Unprecedented also, the amino terminal and carboxyl terminal amino acids are linked covalently in the antibiotic, producing a cyclic structure (Fig. 7.12).

*Penicillin was the first antibiotic known to contain the unnatural amino acid D-alanine.

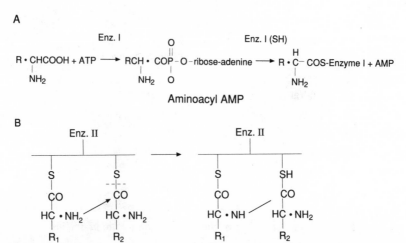

Fig. 7.13 A. Amino acid activation in the synthesis of bacterial peptide antibiotics, a variant of protein synthesis. B. In gramicidin synthesis, enzyme
protein rather than RNA determines the amino acid sequence.

Following the elucidation of gramicidin's structure, the question arose
whether the mode of linking the several amino acids resembled that for
the small model peptide glutathione, or whether it was a variant of ribo-
some (aminoacyl t-RNA) linked peptide formation in protein synthesis.
Neither mechanism turned out to be relevant, except that in all three
systems ATP serves ultimately to activate the amino acid carboxyl group.

In gramicidin synthesis, the initial activation product is aminoacyl AMP,
the same as in protein synthesis, but there the pathways diverge. Two
large enzymes participate, but without the aid of the ribosomal machinery.
Enzyme I catalyzes the activation reaction, as well as the transfer of the
activated amino acid to an adjacent thiol (SH) group of the same protein
(Fig. 7.13A). One energy-rich bond is exchanged for another, and the
product becomes enzyme bound. To add to the complexity, this product
of enzyme I is transferred next to thiol (SH) groups of enzyme II, a protein
that contains ten separate amino acid–specific sites. These are lined up
for linking the precursor units in the correct order, dictated by the amino
acid sequence of the final product (Fig. 7.13 B).

The differences from protein synthesis are profound, both mechanistically and in the mode of information transfer. In gramicidin biosynthesis,

enzyme-bound thioesters are the ultimate reactants for peptide bond formation, in contrast to the ribose-O-esters in ribosomal protein synthesis. Equally striking, the gramicidin-forming enzymes themselves seem to possess the information for coding, selecting the proper amino acids for their alignment in the specified sequence. In a sense, the synthesis of gramicidin and similar bacterial peptide antibiotics is at odds with the dogma that assigns this role to messenger RNA. Eukaryotic cells appear to have abandoned it, or never invented it.

It has been estimated that polypeptide-synthesizing systems of the gramicidin type are probable limited to chains no more than twenty amino acids in length. They function as selective ion carriers, of potassium versus sodium, and owe their antibiotic effects to this property.

Biochemists pondered a related question when a number of hormones were shown to be peptides of varying lengths. This category includes ACTH (adrenocorticotropic hormone), oxytocin, vasopressin, endorphins, enkaphalins, and many others ranging in length from five to several hundred amino acid residues. What mechanisms operate in their synthesis? Once more, the answer was unanticipated. As first shown by Donald Steiner and colleagues for insulin (1967), polypeptide hormones are derived from very much larger proteins, synthesized in eukaryotes by the universal ribosomal machinery. These polyproteins are biologically inactive (prohormones or preprohormones). Proteolytic enzymes cleave the high-molecular-weight precursor at selected sites to smaller, hormonally active fragments. Such peptide hormones are catabolic products of a special kind, in turn controlled by unknown signals.

To chemists of earlier generations, the process of peptide bond formation appeared to be rather simple and straightforward. That Nature chose not only one but at least three different chemical mechanisms, none of them foreseen, should teach scientists a lesson. Chemical arguments for biological processes can address only the possible, not the actual (Jacob, 1982). Even after we understand a given reaction mechanism, there is still the question why Nature chose it—trial and error?

Bibliography

Fatty Acid Metabolism

1. F. Lynen (1964), The road from "activated acetic acid" to the terpenes and the fatty acids, *Les Prix Nobel,* Nobel Foundation, pp. 205–244. Stockholm, Imprimerie Royale, P. A. Norstedt.

2. S. Wakil (1959), A malonic acid derivative as an Intermediate in fatty acid synthesis, *J. Am. Chem. Soc.* **80,** 6465.

3. R. O. Brady (1958), The enzymatic synthesis of fatty acids by malonic acid condensation, *Proc. Nat. Acad. Sci.* (USA) **44,** 993.

Glycogen Synthesis and Breakdown

4. C. F. Cori (1946), Enzymatic reactions in carbohydrate metabolism, *The Harvey lectures,* pp. 253–272. New York, Academic Press.

5. C. F. Cori, G. Schmidt, and G. T. Cori (1939), Synthesis of a Polysaccharide from glucose-1-phosphate by muscle extract, *Science* **89,** 464.

6. L. Leloir (1970), Two decades of research on the biosynthesis of saccharides, *Lex Prix Nobel,* Nobel Foundation, pp. 178–188. Stockholm, Imprimerie Royale, P. A. Norstedt.

7. J. T. Park and M. J. Johnson (1949), Accumulation of labile phosphate in *Staphylococcus aureus* grown in the presence of penicillin, *J. Biol. Chem.* **170,** 585.

Peptide and Protein Synthesis

8. J. T. Park and J. Strominger (1957), Mode of action of penicillin: Biochemical basis for mode of action and for its selectivity, *Science* **125,** 191.

9. J. Snoke and K. Bloch (1954) The biosynthesis of glutathione, in *Glutathione,* pp. 129–137. New York, Academic Press.

10. F. Lipmann and L. C. Tuttle (1945), A specific micromethod for the determination of acylphosphates, *J. Biol. Chem.* **159,** 21.

11. N. Bucher (1953), The formation of radioactive cholesterol and fatty acids from C^{14}-labeled acetate by rat liver homogenates, *J. Am. Chem. Soc.* **75,** 498.

12. P. C. Zamecnik (1959), Historical and current aspects of the problem of protein synthesis, *The Harvey lectures,* pp. 257–281. New York, Academic Press.

13. M. Hoagland (1955), An enzyme mechanism for amino acid activation, *Biochem. Biophys. Acta* **16,** 288.

14. D. F. Steiner et al. (1967), Insulin biosynthesis: Evidence for a precursor, *Science* **159**, 697–700.

15. W. Jencks (1969), *Catalysis in chemistry and enzymology,* pp. 324–325. New York, Academic Press.

16. W. Zachau and W. Karau (1960), Reaktionsfähige Aminosäure ester, *Chem. Ber.* **93**, 180.

17. W. Gilbert (1986), The RNA world, *Nature* **319**, 618.

18. L. Orgel (1989), The origin of polynucleotide directed protein synthesis, *J. Mol. Evol.* **29**, 465–474.

19. F. Lipmann (1971), Attempts to map a process of evolution of peptide biosynthesis, *Science* **173**, 875.

20. A. Eschenmoser et al. (1990), Aldolmisierung von Glycolaldehyde zu Ribose 2,4-diphosphate, *Helv. Chim. Acta* **73**, 1410–68.

21. Y. Ovchinnikov and V. Ivanov (1982), *The cyclic peptides: Structure, conformation and function,* vol. 5, p. 547. New York, Academic Press.

22. F. Jacob (1982), *The possible and the actual,* p. 25. New York, Pantheon Books.

8

Alternative Pathways

———

Here we consider profligacy versus parsimony in biological

systems. Alternative pathways to identical body

constituents in several instances. Why?

"In the whole drama of nature we find waste and prodigality."* "Selfish genes," "junk DNA," and a few instances of duplicate pathways to be discussed in this chapter would appear to support the above statement. At variance is the notion of parsimony, the economy of specific means to an end, an element of the wisdom of Nature and implicit in Darwinian evolution. The sum total of biochemical experience to date, the universal pathways producing the majority of cell constituents, certainly suggests parsimony as the dominant principle. Though rare, examples of profligacy do exist. We will examine some of the examples of alternative pathways, whether they constitute a squandering of cellular resources, a duplication of effort, or whether there is some underlying wisdom. Perhaps genetic manipulation will provide an answer.

In the definition used here, alternative pathways are those that occur *side by side in the same organism.* They are not replacements of one by another, that is, chemically distinct starting from different precursors but

*John Burroughs, American naturalist, 1837–1921.

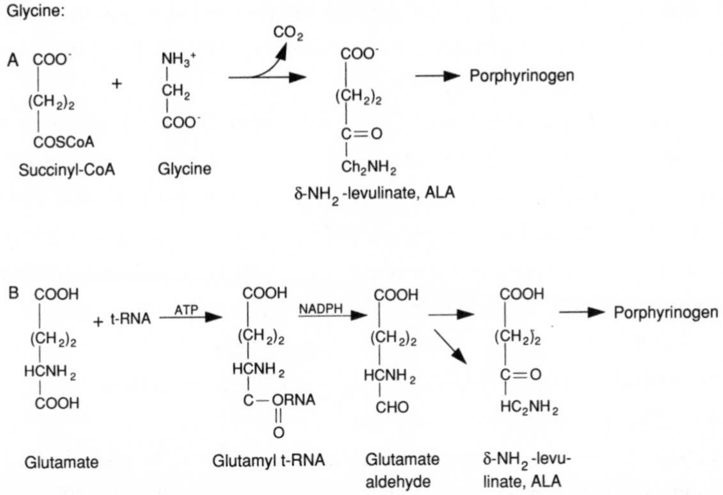

Glycine:

A
$$CO_2$$

COO^-		NH_3^+		COO^-	
$(CH_2)_2$	$+$	CH_2	$\longrightarrow$	$(CH_2)_2$	$\longrightarrow$ Porphyrinogen
$COSCoA$		COO^-		$C=O$	
				Ch_2NH_2	

Succinyl-CoA   Glycine δ-NH_2-levulinate, ALA

B

$COOH$		$COOH$		$COOH$		$COOH$	
$(CH_2)_2$	$+$ t-RNA $\xrightarrow{ATP}$	$(CH_2)_2$	$\xrightarrow{NADPH}$	$(CH_2)_2$	$\searrow$	$(CH_2)_2$	$\longrightarrow$ Porphyrinogen
$HCNH_2$		$HCNH_2$		$HCNH_2$		$C=O$	
$COOH$		$C-ORNA$		CHO		HC_2NH_2	
		$\parallel$					
		O					

Glutamate   Glutamyl t-RNA   Glutamate δ-NH_2-levu-
                                       aldehyde   linate, ALA

Fig. 8.1 The Shemin pathway (A) for the formation of delta-aminolevulinic acid (ALA), the source of all carbon and nitrogen atoms of pyrrols; and (B), the alternative pathway to ALA by way of glutamyl t-RNA.

leading to the same end product. The most clear-cut case of one pathway *replacing* another is the biosynthesis of monounsaturated fatty acids (see Chapter 3). In this instance the appearance of oxygen in the atmosphere was the driving force for replacement. An anaerobic pathway employed by many bacteria was superseded by an aerobic mechanism, without any vestiges of the former remaining. The case is one of substitution, not redundance.

Delta-Aminolevulinic Acid

Three examples of nature's apparent profligacy, and possibly a fourth, come to mind. The first is the existence of two independent pathways for δ-aminolevulinic acid, the common and universal precursor for the porphyrin moiety of heme, of chlorophyll, and of vitamin B_{12} (Fig. 8.1).

The mechanism discovered by David Shemin in 1945, the only one known for three decades, was in a sense the outcome of serendipitous observations made in the course of experiments designed to measure the life span of the red cell. His findings on this subject led to important information, but also provided him with an entry into the still unknown mechanism of porphyrin biosynthesis.

The tracer amino acid* was ^{15}N glycine and Shemin himself was the experimental animal. His blood was withdrawn periodically and the ^{15}N content of hemoglobin determined in successive blood samples. From its decline, the survival time or life span of the human red cell was calculated and found to be approximately one hundred days. In addition to the ^{15}N concentrations of the hemoglobin, Shemin's red cell protein and the heme component of hemoglobin were also analyzed for ^{15}N. Glycine proved to be the nitrogenous precursor of heme's porphyrin structure.

How Shemin discovered succinic acid as the carbon source for the porphyrin's 4-pyrrole rings is a more complex story. The logical reasoning of a seasoned biochemist and ingenious interpretation of the data clarified the labeling pattern of the porphyrin molecule, culminating in the postulate that the previously unknown compound δ-aminolevulinic acid (ALA) formed from glycine and succinate was the immediate precursor of the porphyrin structure (Fig. 8.1A). Thus all carbon and nitrogen atoms of this complex molecule were accounted for.

To investigate and demonstrate the glycine + succinate → δ-aminole-vulinate pathway, Shemin chose animal cells. Immature avian erythro-cytes proved to be the richest enzyme source. Later, a survey of chloro-phyll synthesis in algae and higher plants revealed—surprisingly—a chemically novel mechanism for porphyrin synthesis. It started with glu-tamic acid, the sole carbon and nitrogen source for the plant porphyrins, instead of succinate and glycine (Fig. 8.1B; Beale and Castelfranco, 1973).

Pathways A and B to ALA bear no mechanistic resemblance whatsoever to each other. Also, A is catalyzed by a single mitochondrial enzyme; B, by a multienzyme complex located in plant chloroplasts. Because light intensity controls the greening of plants, it is no surprise that the initial step of chlorophyll synthesis, δ-aminolevulinate synthesis from glutamate, is light controlled. By contrast, the first step in porphyrin synthesis in

---

*It should be emphasized that the majority of the twenty amino acids, the protein building stones, are employed in specific metabolic pathways separate from their role as determinants of protein structure and function.

animal blood—from glycine and succinate—is regulated by heme, the ultimate product in mitochondria, an example of a phenomenon known as end product or feedback control. Most prominent among the mechanistic differences is the unexpected participation of transfer or t-RNA in the "plant" or glutamate pathway, reported by several laboratories between 1984 and 1986.* The clue came from a routine experiment, showing that ALA synthesis in plants is abolished by the RNA-cleavage enzyme ribonuclease.

We have already encountered transfer or t-RNA as the carrier of activated amino acids in peptide and protein synthesis, its only previously recognized function. As shown in pathway B, glutamyl (aminoacyl) t-RNA serves as the substrate for reducing the activated glutamate carboxyl group to an aldehyde.** Probably, but not yet established conclusively, it is the same glutamyl t-RNA that provides the activated amino acid for protein synthesis.***

The distribution or occurrence of the two separate pathways to ALA among organisms is shown in Table 8.1.

Why organisms have chosen one or the other pathway to porphyrin precursors is not immediately obvious. Aerobic versus anaerobic lifestyles of prokaryotic and eukaryotic cells, or for that matter of plants and animals, do not seem to be determinants. One can make a strong case that ALA synthesis from glutamate is the pathway chosen by the plant kingdom and destined ultimately for the synthesis of chlorophyll, rather than porphyrin synthesis for hemoglobin or the cytochromes involved in respiration. Yet this classification breaks down in several instances. Tubers of plants, in contrast to plant leaves, contain the enzymes catalyzing the dark process, ALA synthesis from succinate and glycine (pathway A). Therefore the same organism harbors two independent sets of genes for producing the identical porphyrin precursor, albeit in morphologically

*Classic biochemistry was declared "dead" about a decade ago, a pronouncement obviously premature.

**There are other examples of reducing RCOOH to RCHO, but in none of them does RNA intervene.

***Regrettably, this novel role for t-RNA is not mentioned in modern, otherwise comprehensive texts of biochemistry (with one exception).

Table 8.1
Distribution of the two pathways to δ-aminolevulinate (ALA)

ALA from glutamate[a]	ALA from succinate and glycine[b]
Higher plants	Tubers of higher plants: potatoes,
Tomatoes	nonphotosynthetic plant tissues
Ferns, mosses	
Algae	
Euglena gracilis (green)	*Euglena gracilis* (etiolated)
Photosynthetic bacteria	
Archaebacteria	Bacteria, both anaerobic and aerobic
Fungi	Yeasts
Animal tissues?	Animal mitochondria

Source: C. G. Kanangara, et al., *Trends Biochem. Sci.,* April 13, 1988.
a. Reactions that are light dependent are shown in italics.
b. All reactions occur in the dark.

separate locations or different subcellular compartments. Equally striking, *Euglena gracilis,* a unicellular protozoan* or "algal flagellate," practices a lifestyle that is a hybrid of plant and animal cells. When grown in the light, photosynthesis is its energy source. When green *Euglena* cells are adapted to growth in the dark, they bleach, lose their chloroplast (become etiolated), and rely on organic nutrients, a heterotrophic energy source. It therefore became crucial to determine whether or not the two pathways to ALA occur side by side in green and etiolated *Euglena.* The evidence seems clear that the light-dependent glutamate pathway predominates in green *Euglena.* In etiolated cells of the same algae flagellate (those that lack chloroplasts) the succinate plus glycine route to ALA, the dark mechanism, prevails but the alternative pathway is not entirely absent.

Known to every gardener, dark-grown seedlings of higher plants develop yellow leaves for lack of chlorophyll. When bathed in solutions containing ALA, the etiolated leaves turn green. this phenomenon of course remained

*André Lwoff considers it impossible to give a satisfactory classification of protozoans.

unexplained prior to discovery of the light-dependent formation of ALA by the glutamate pathway B.

The occurrence of the "plant" pathway (for example, in fungi, a number of heterotrophic bacteria, and also in animal tissues—all nonphotosynthetic systems; see Table 8.1) presents another serious predicament, quite apart from the phenomenon that two independent routes leading to identical products occur side by side in the same organism. I suspect the duality and distribution of the pathways to δ-aminolevulinate raise problems for the evolutionary biologist.

Dual Pathways to Phosphatidylcholine

We have addressed the biosynthesis of phosphatidylcholine (PC), the principal membrane phospholipid, in the context of "The Importance of Being Contaminated" (Chapter 4), the example being the chance discovery in 1956 that CTP (cytidinetriphosphate) in addition to ATP is an essential cofactor for choline activation in animal tissues. Somewhat later, a second independent pathway to this membrane phospholipid appeared on the scene (Bremmer and Greenberg, 1961).

In this pathway, in contrast to the one discovered by Kennedy (Chapter 4), phosphatidylethanolamine (PE), another common membrane phospholipid, is converted to PC by sequential transfers of three methyl groups to an amino acceptor (Fig. 8.2). The fact that the two independent pathways to chemically identical products occur side by side in the same organism, for example yeast or animal tissues, appears to be the clearest example of redundancy. Why should cells employ two independent processes, one requiring one or at the most two enzymes (methylation pathway) and another that involves at least twice as many protein catalysts, to achieve the same end result?

Mutants deficient in enzymes for one of the pathways might provide some clues. If such mutants grow at the same rate as the corresponding wild type, then one of the pathways would be expendable. Some of my laboratory findings suggest but do not prove that this might be the case. Exposure of yeast to specific inhibitors of the Kennedy pathway or that shown in Fig. 8.2 stimulates the alternative pathway, with the net result

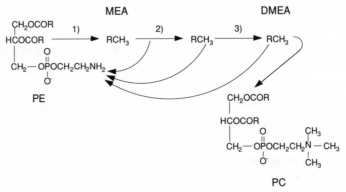

Fig. 8.2 The alternative pathway of phosphatidylcholine synthesis by methylation of phosphatidylethanolamine. A single enzyme catalyzes the three
methylation steps.

that the total content of cellular phosphatidylcholine remains nearly the
same as in uninhibited cells. Given that the cell compensates for declining
rates of one of the pathways by raising the rate of the alternative route,
then the products of the two must indeed be functionally identical. Perhaps
the phenomenon is a safeguard, protecting the organism in stressful situations that have yet to be examined.

The existence of two independent routes to the same end product might
also be rationalized if an intermediate of one of the pathways lies at a
branch point, serving as an essential precursor for more than one end
product (Fig. 8.3). However, such branch points are not known for either
of the two routes leading to the biosynthesis of phosphatidylcholine.

Is it possible to date the appearance of one or the other pathway in the
course of evolution? In a few bacterial membranes, for example that of

Fig. 8.3 A possible rationalization for alternative pathways leading to identical products by branching.

*Agrobacterium tumefaciens,** phosphatidylcholine occurs; and when it does, the phospholipid precursors, monomethyl and dimethyl phosphatidylethanolamines (MEA and DMEA, Fig. 8.2) are found in the bacterial membrane along with phosphatidylcholine. It therefore appears that the methylation pathway (2) starting with phosphatidylethanolamine proceeded to completion in only a few bacterial species.

One morphological clue to the origin of the methylation pathway has been noted (Hagen and Goldfine, 1966). Partial or complete biosynthesis of phosphatidylcholine occurs in those bacteria that have developed extensive internal membranes (such as *Hyphomicrobium*), a trait rare in prokaryotes but universal in eukaryotic cells. Clearly, methylation of phosphatidylethanolamine evolved in stages, suggesting that this pathway was the earlier (more primitive?) of the two. Comparative studies also reveal the total absence of the more "expensive" pathway 1 (of Kennedy) in any of the bacteria studied. One should add the qualifier *organisms examined so far,* because occasionally creatures of the microbial world render generalizations invalid.

Evolution of C_3 + C_4 Photosynthesis

The purist may argue that *sensu strictu* the next example does not qualify as a case of alternative pathways. Admittedly one of these pathways is merely an adjunct to the other, but it does not operate independently in the photosynthetic process. It is more a variation on a theme, making photosynthesis more efficient under certain circumstances.

Until the early 1960s the famed Calvin-Benson cycle for photosynthesis was thought to be the only device used by nature for fixation of atmospheric carbon dioxide by green plants, resulting ultimately in the synthesis of carbohydrates (glucose or starch). Schematically, the Calvin cycle can be formulated as shown in reaction (1) of Fig. 8.4: carbon dioxide attaches to C_2 of a five-carbon sugar, the resulting carboxy derivative

*These are bacteria infecting various higher plants, causing crown gail tumors.

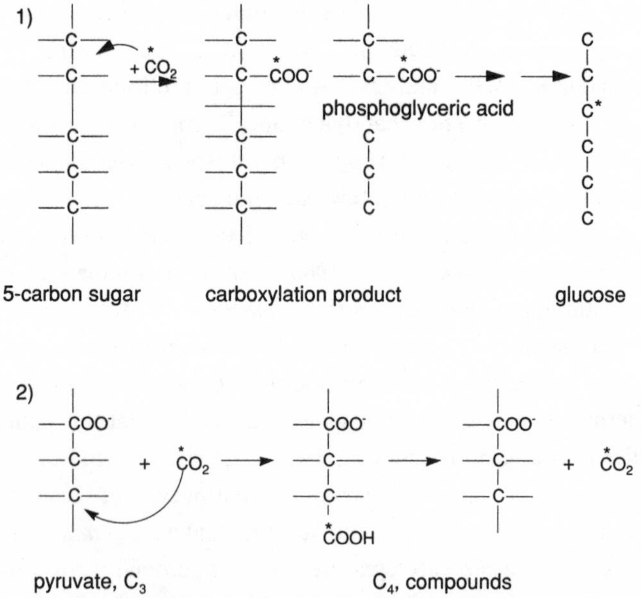

Fig. 8.4 Mechanisms for (1) CO_2 fixation in green plants and algae, and (2) CO_2 fixation into C_4 compounds, serving to increase intracellular CO_2 pressure for reaction (1).

undergoing cleavage to two molecules of the three-carbon compound phosphoglyceric acid, in turn the precursor of glucose. This is the principal mechanism in the algae and spinach leaves studied by M. Calvin and A. A. Benson (1945–1953). The choice of these plant systems was crucial, as we shall see. These studies were among the earliest to owe their spectacular success to the use of $^{14}CO_2$ as a tracer.

When leaves of flowering plants (angiosperms) were exposed *very briefly* to $^{14}CO_2$, however, the radioactive carbon appeared earliest in a number of four-carbon compounds (oxaloacetate, malate, or aspartate)—hence "C_4" plants—while in "C_3" plants the first detectable radioactivity resides predominantly in the three-carbon compound phosphoglyceric acid.

In principle, these C_4 compounds can be converted to carbohydrates, but for complex reasons without net gain. This cycle by itself would be futile because the fixed CO_2 is lost again, as seen in reaction (2) of Fig. 8.4. It is now known that the sole function of C_4 synthesis is to "feed" and raise the concentration of CO_2 entering the Calvin cycle, which is located in a separate cell compartment.

The following hypothesis has been proposed and verified. In higher plants the "productive" CO_2 fixation into C_3 compounds by way of C_5 (Fig. 8.4, reaction 1) takes place *internally* in the so-called bundle sheath cells, while the more rapid CO_2 fixation, $C_3 + CO_2 \rightarrow C_4$ (reaction 2) occurs in *mesophyll* cells directly exposed to air and external to the bundle sheath cells. This process is faster and more effective because the mesophyll receptor system has a much higher affinity of binding capacity for CO_2 than the bundle sheath cells. The CO_2 released from the C_4-CO_2 fixation product is not wasted but passes into the bundle sheath cells, enters the Calvin cycle, and renders it more efficient due to increased local CO_2 pressure. Strictly speaking, the C_4 pathway is an auxiliary device, incapable on its own of productive CO_2 fixation and hence of carbohydrate synthesis. By contrast, the Calvin cycle is self-sufficient in many organisms.

To classify photosynthetic organisms based on the two types of photosynthetic CO_2 fixation, an appropriate distinction would be between C_3 cells, which are autonomous, and C_3-C_4 plants, which refer to the obligate cooperative CO_2 fixation. "Pure" C_4 plants lacking the Calvin cycle do not exist. In essence, anatomical and physiological traits are responsible for the classification of C_3 and C_3-C_4 photosynthetic organisms. Typically, the operative sites of Calvin cycle reactions are associated with organelles that include chloroplasts and mitochondria residing in higher, leaf-bearing plants. In bacteria and unicellular algae such as *Chlorella* (the cells used in Calvin's work), CO_2 supply feeding the cycle that bears his name is apparently not limiting. His choice of organisms could be one of the reasons why the CO_2 fixation into C_4 dicarboxylic acids was minimal or, more likely, not detected at early time points.

As for the distribution of the C_4 mechanism for primary CO_2 fixation coupled to the Calvin cycle ($C_3 + C_4$), the hot climates and high light intensities prevailing in the tropics and in deserts provide favorable environments. To repeat, strictly speaking the pathway is an auxiliary, not a chemical alternative. As for the phyletic history of the C_3 and the $C_3 + C_4$ photosynthetic systems, "the answer will probably depend on where one looks for it" (Moore, 1982).

Pathways to Nicotinamide

All forms of life, aerobes and anaerobes, require this universal molecule as a component of the pyridine nucleotides, the coenzymes NAD and NADP (or their reduced forms NADH and NADPH). The inclusion of nicotinamide under the heading of alternative pathways is arbitrary. It might have been discussed, perhaps more appropriately, in the context of "Oxygen and Evolution" in chapter 3—the reason being that one pathway is oxygen dependent.

As noted before, the biosynthesis of nicotinamide proceeds by oxidative degradation of the amino acid tryptophan, confining this process to respiring, or aerobic, cells. Shortly after the tryptophan → nicotinamide transformation was elucidated and found to be oxygen dependent at three early stages, attention turned to an exploration of pathways that should necessarily proceed in anaerobic microorganisms.

Inspection of the multistep tryptophan → nicotinamide conversion locates the three oxygen-dependent steps between tryptophan and quinolinic acid (Fig. 8.5). Oxygenase enzymes are not involved in the further transformation of this intermediate to nicotinamide. One might therefore conjecture that in any putative anaerobic path to nicotinamide, quinolinic might also serve as an intermediate.

O. Hayaishi, a pioneer in the discovery of oxygenases* and the principal investigator of tryptophan metabolism, thought along these lines. He reasoned that quinolinic acid** is also a likely candidate for an anaerobic path to nicotine amide. Presumably Hayaishi was aware of earlier findings (Ortega and Brown, 1960) that in bacteria tryptophan did not serve as a precursor to nicotinamide. In crucial experiments in 1963 Hayaishi's laboratory showed that *E. coli* was indeed able to convert quinolinic acid, a product of tryptophan metabolism, to the nicotinamide-containing ribonucleotide (NAD).

---

*H. S. Mason (1955) independently discovered oxygenases, enzymes that catalyze the entry of oxygen into organic molecules.

**Quinolinic acid per se appears to have no other metabolic function. It is listed as having neuroexcitatory activity and possibly a role in degenerative brain disorders.

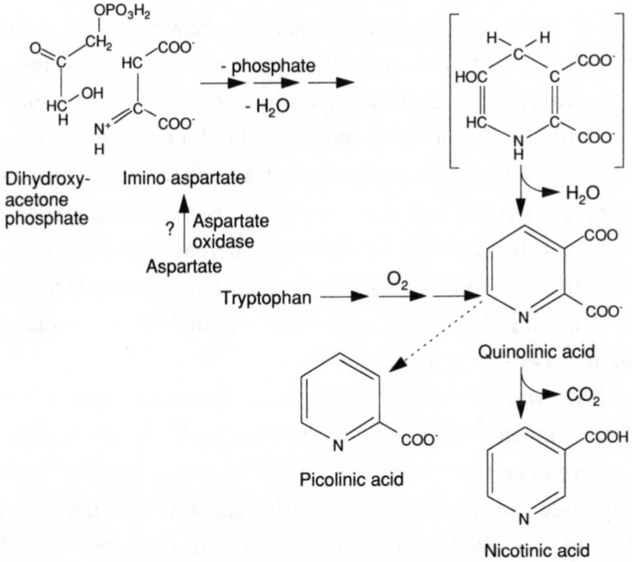

Fig. 8.5 An anaerobic pathway leading to quinolinic acid, a precursor of nicotinamide (trade name Niacin). In nicotinamide $CONH_2$ replaces the COOH group of nicotinic acid.

As for the origin of quinolinic from precursors other than tryptophan, evidence obtained in various laboratories pointed to a de novo synthesis, from small molecules commonly available in intermediary metabolism, not degradation of a large precursor. The small molecules were identified as aspartic acid (an amino acid) and dihydroxyacetone phosphate, an intermediate of carbohydrate metabolism (see Fig. 8.5).

Mechanistic details of the pathway are tentative except for labeling studies which show that dihydroxyacetone phosphate and aspartic acid provide the seven carbon atoms and the nitrogen of the pyridine ring of quinolinic acid. Genetic studies suggest that no more than two enzymes are involved, observations that do not exclude a multistep process.

The first of the two enzyme activities, known as aspartate oxidase, converts the amino acid initially to the imino derivative, a reactive and therefore unstable intermediate that cannot be isolated. Reasonable reaction mechanisms for the following series of events can be written, but we have no proof. A major problem, still unresolved, concerns the first step proposed by all investigators in the field, the aspartic → iminoaspartic acid conversion. The *E. coli* enzyme, known as aspartate oxidase, implies an oxidative reaction. Indeed, assayed in vitro, *E. coli* aspartate oxidase has an absolute requirement for oxygen. The dilemma arises because the same pathway is referred to in the literature as anaerobic. Thus a mutation for the gene coding for aspartate "oxidase" is expressed under both anaerobic and aerobic growth conditions, whereas in vitro oxygen is the obligate electron acceptor. Also, *Clostridium butylicum,* a strict anaerobe, is said to form quinolinic acid from aspartate (or N-formylaspartate?) and an unspecified two-carbon compound.

Research on this alternative pathway to quinolinic acid and hence to nicotinic acid appears to have been abandoned during the last decade. At any rate, there is no doubt that a relatively simple (primitive?) mechanism for nicotinamide synthesis exists in microorganisms, one that is apparently retained in higher plants. The "reason" for its existence may be quite unrelated to aerobic or anaerobic lifestyle. More probably, anaerobic bacteria have not yet invented the enzymatic machinery for degrading tryptophan to quinolinic acid.

Whether or not the two pathways to nicotinamide coexist in the same organism—as they do in the syntheses of phosphatidylcholine and delta-aminolevulinic acid—remains to be seen. The answer appears to be a qualified yes. The common yeast *Saccharomyces cerevisiae,* when grown aerobically, uses the degradative tryptophan pathway; it uses the de novo pathway from small molecules during anaerobic growth. Still, it appears that in eukaryotes tryptophan degradation is the dominant if not exclusive route. The question remains why most higher organisms, specifically animals, have abandoned the biosynthetic or de novo route to nicotine amide from aspartate and dihydroxyacetone phosphate in spite of its apparent simplicity.

Postscript

Too thorough a literature search may at times disclose relevant, though not necessarily essential information. In an earlier version of this chapter I omitted mentioning an alternative pathway of tryptophan degradation to an isomer of nicotinic acid named picolinic acid (Fig. 8.5). This molecule fails to replace nicotinic acid nutritionally. In cat liver the diversionary picolinate-forming enzyme is extraordinarily active, accounting for the niacin requirement and—inter alia—for the stringent nutritional requirements of the carnivorous felines (see Chapter 12).

Bibliography

Formation of the Heme Precursor

1. D. Shemin and C. Russell (1953), 5-Aminolevulinic acid, its role in the biosynthesis of porphyrins, *J. Am. Chem. Soc.* **75**, 4583.

2. D. Shemin (1955), The biosynthesis of porphyrins, *The Harvey lectures,* p. 258. New York, Academic Press.

3. S. Beale and P. Castelfranco (1973), ^{14}C incorporation from exogenous compounds into δ-aminolevulinic acid by greening cucumber cotyledons, *Biochem. Biophys. Res. Comm.* **52**, 143.

4. D. D. Huang el al. (1984), δ-Aminolevulinic acid synthesizing enzymes need an RNA moiety for activity, *Science* **225**, 1482.

5. J. Weinstein and S. Beale (1983), Separate roles for two tetrapyrrole biosynthetic pathways in *Euglena gracilis, J. Biol. Chem.* **258**, 6799.

6. C. Garmini et al. (1988), t-RNA glu as a cofactor in δ-aminolevulinate synthesis, *Trends Biochem. Sci.* **13**, 139.

7. S. Beale and J. Weinstein (1991), Biosynthesis of 5-aminolevulinic acid in phototrophic organisms, *Chlorophyll* **385**.

Phospholipid Biosynthesis

8. F. Bremmer and D. Greenberg (1961), Methyl-transferring enzymes in the biosynthesis of lecithin, *Biochim. Biophys. Acta* **46**, 205.

9. P. Hagen and H. Goldfine (1966), Phospholipids of bacteria with extensive intracytoplasmic membranes, *Science* **151**, 1543.

C_3 and C_4 Photosynthesis

10. M. Calvin (1951), The path of carbon in photosynthesis, *The Harvey lectures,* p. 218. New York, Academic Press.

11. D. Youvan and B. Marrs (1989), Molecular mechanism of photosynthesis, *Sci. Am.* **256**, 42.

12. P. Moore (1982), Evolution of photosynthetic pathways in flowering plants, *Nature* **295**, 647.

Nicotinic Acid (Niacin)

13. H. White (1982), *Biosynthetic and salvage pathways of pyridine nucleotide coenzymes,* p. 225. New York, Academic Press.

14. S. Nakamura et. al. (1963), A mechanism of niacin-ribonucleotide formation from quinolinic acid, *Biochem. Biophys. Res. Comm.* **13**, 285.

15. O. Hayaishi et al. (1967), Studies on the biosynthesis of NAD by a direct method of tracing metabolism *in vivo, Adv. Enz. Reg.* **5**, 9–22.

16. M. V. Ortega and G. M. Brown (1960), Precursors of nicotinic acid in *Escherichia coli, J. Biol. Chem.* **235**, 2939–45.

9

Lactose Intolerance

———

A widespread metabolic disorder in adults is caused by

defective digestion of the milk sugar lactose. Practices for

avoiding this disorder are discussed.

"Many people cannot drink milk without becoming ill. It is estimated that more than 70 percent of the world's population are unable to tolerate milk sugar (lactose) because they lack an enzyme (lactase) needed to help break down the milk sugar for proper digestion" (*MGH News,* **49**[1990], 7).

My interest in this intriguing subject dates to my student days, more than fifty years ago. I spent the early 1930s at the Technische Hochschule in Munich studying chemistry. Occasionally a family friend, the owner of a brewery, would invite me for Sunday dinner. The conversations during these welcome occasions were lively.

I remember one in particular. Rumors had reached my host that a growing number of university students had taken to drinking milk with their meals. Was this rumor true, he wanted to know. I had to admit that my observations confirmed it, attributing the unmanly habit to the influx of foreign students, mostly from the United States. "Let me tell you," the brewer said, "Nature intended milk to be food for infants, not for adults. In fact, all mammalian species stop drinking milk after weaning. Instinct tells them that it is not good for their health."*

———

*My friend—although he did not mention it—may have read a very early

By the end of the conversation he had persuaded and commissioned me to search the university library for evidence and arguments that might help him halt this threat to the brewing industry. My literature search yielded little of substance, apart from some clinical evidence. Infants were believed to digest milk with the aid of intestinal enzymes more efficiently than adults.

The brief paper I managed to prepare satisfied my friend and brought this impecunious student the generous reward of 100 marks. It was most welcome, arriving as it did during *Fasching,* the Bavarian equivalent of Mardi Gras, when students were down to their last penny. Today I could argue the brewer's case with sounder evidence. Milk is not necessarily conducive to health—in adolescents or adults.

Consumption of Milk and Dairy Products

A visitor to college dining halls in the United States, especially one from abroad, tends to be intrigued by the sight of a battery of glasses of milk on a student's tray. No doubt this practice is or was deemed superior nutritionally to cola drinks. Certainly it is no longer compatible with the current advice to limit dietary fat to 30 percent of caloric intake. Fortunately, low-fat milk is rapidly replacing the "whole-milk" variety, and without any loss of nutritional quality. Why has milk been or become so popular among our young people? Perhaps technological advances early in the twentieth century played a major role. Pasteurization, refrigeration, and homogenization made milk less perishable and therefore more palatable to all age groups. Still, extraordinarily large milk consumption seems to be a North American phenomenon.

Some Consequences of Milk Intolerance in Young Adults

Interesting and perhaps alarming postmortem studies have been conducted on American soldiers, combat casualties during the Korean and

---

paper stating that "cows have a much smaller milk sugar digestion than calves" (Roehmann and Lappe, 1895).

Vietnam wars. In Korea, of three hundred soldiers killed, age twenty two on average and all Caucasians, 77 percent had gross evidence of incipient coronary heart disease. In a similar study of Japanese natives of the same age group but not soliders, less than 2 percent showed such symptoms. Combat stress was therefore considered a likely cause of these remarkable differences. Dietary habits were also mentioned as a possible factor. It should be noted that the Korean War was fought by United Nations troops, including Australian soldiers. I remember—but have no documentary proof—that soldiers from "down under" showed no postmortem evidence of incipient cardiovascular pathologies. Is it a reasonable guess that beer, not milk, was the preferred beverage of the Australian combatants?

A study conducted on U.S. soldiers eighteen years later during the Vietnam War showed a significant decline in atherosclerotic symptoms, from the 77 percent of the Korean sample to 45 percent. I would speculate that in the interim milk either became less popular in the States or was replaced by low-fat milk.

Why the popularity of milk among young adults has not spread throughout the Western world may have a variety of reasons. Until quite recently, Old World countries have been more conservative, retaining their traditional regional choices for food and drink. Statistical surveys provide some relevant but incomplete information. If we compare, for example, the daily per capita milk supply in the countries of western Europe, the range is relatively small (Table 9.1). It varies from a low of 300 gm in Italy to a high of 502 gm in Sweden. The values for the United States and Switzerland are essentially identical (about 460 gm). Yet "milk supply" or production does not necessarily mean consumption of fluid milk as such. A substantial but varying portion goes into dairy products such as butter, cheese, ice cream,* and the like.

*In the 1980s a headline in the *New York Times* announced what it called "Buttergate in Vermont." A cooperative creamery in the northern part of the state labeled its dairy products as Vermont butter, Vermont cheese, and so on. As the Federal Trade commission discovered, the creamery, in order to meet demand, had been forced to import milk from neighboring states because Vermont milk was in short supply. There were two reasons for the shortage: the federal program to buy up dairy farms, and the establishment of a Ben &

Table 9.1

Per capita supply of milk and percentage of population with lactase persistence

Country	Per capita milk supply (gm/day)	Lactase persistence (%)
Sweden	502	99
Denmark	432	97
United Kingdom (Scotland)	455	95
Germany	329	88
Switzerland	461	84
Australia	331	82
United States (Iowa)	462	81
Spain	329	72
France	347	58
Italy	300	49
India	105	36
Japan	135	10
China (Shanghai)	13	8
China (Singapore)	113	0

Sources: Percentages from United Nations balance sheets, 1979–1981, quoted in Cramer (1989).

For example, with the same per capita milk supply, Switzerland produces four times as much cheese as the United States and Canada. The importance of this issue will become apparent from our later discussion of milk intolerance.

Dietary Laws

No information exists on whether in biblical times milk intolerance was common among Palestine's Hebrew population. In present-day Israel, the incidence is very similar to that prevailing in countries around the Medi-

---

Jerry's ice cream production plant, which accounted for 25 percent of the dairy state's residual milk production. The cooperative was forced to change its label to "New England's Finest Creamery Product."

terranean rim. At any rate, milk was obviously consumed in biblical times because kosher diets, sanctioned by Jewish law, prohibit the simultaneous preparation and consumption of meals containing meat and milk. With the recent arrival of milk substitutes (see below) Hebrew theologians today face the issue of whether milk substitutes are kosher (fit). Some rabbinates sanction their use (Rabbi Ben Zion Gold, private communication). Also, dairy products such as cheese have been traditionally approved as kosher. Perhaps it was argued that milk products such as casein (the bulk protein component of cheese) no longer conform to the animal liquid defined by Jewish law. That milk intolerance played a role in the dietary restrictions seems doubtful.

Lactose Intolerance

The mother's milk infants receive is a complete source of essential and beneficial nutrients. Science has seen no need to improve on Mother Nature's formula. One potentially troublesome component of this perfect infant food, and produced only by the lactating mammary gland, is the disaccharide lactose, or milk sugar (Fig. 9.1). Infants—and only infants— need it as the sole source of galactose, a sugar widely utilized for a variety of metabolic purposes. Like most disaccharides, lactose is very poorly absorbed from the intestines into the bloodstream, and therefore is first cleaved by the enzyme lactase to galactose and glucose. These two monosaccharides readily enter the circulatory system for further utilization.*

Except in some rare disorders, the offspring of all mammalian species contain lactase from birth, but probably not during early fetal development. There are exceptions to this generalization—as there often are. Pacific sea lion milk contains neither lactose nor the enzyme lactase; the

*Galactose is the characteristic sugar of a family of glycolipids known as gangliosides, found in high concentrations in the gray matter of the nervous system. Faulty ganglioside metabolism, the absence of an enzyme, is the cause of a fatal genetic disorder known as Tay-Sachs disease, associated with blindness and degeneration of the nervous system. Death occurs before the age of ten.

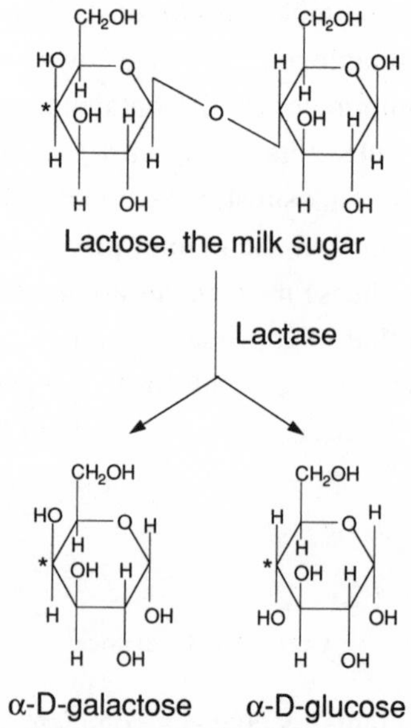

Lactose, the milk sugar

Lactase

α-D-galactose   α-D-glucose

Fig. 9.1 The structure of the disaccharide lactose, converted by the enzyme lactase to galactose and glucose. Note that in the two products the H and OH groups linked to the marked (*) carbon atom have the opposite orientation. The two monosaccharides are stereoisomers, or epimers.

pups are extremely milk and lactose intolerant. Without exception, however, lactase production in mammals first declines and then ceases entirely after weaning, as my brewer friend asserted fifty years ago. Intriguingly, this developmental loss of lactase occurs in some human infants, the lactose intolerant, while others continue to produce the enzyme throughout their lives. Lactase deficiency appears to be inherited as an autosomal recessive trait and in afflicted individuals is associated with abdominal distension, diarrhea, and nausea, and in severe cases produces cataracts. One may speculate that the undigested lactose in the gut causes the bacterial flora to become "unfriendly." ("Hydrogen breath," caused by bacterial fermentation, is one diagnostic tool for detecting lactose intolerance.)

Milk- or lactose-intolerant adults enjoy perfect health, provided they avoid drinking more than a glass or so of milk per day; but they must have an alternative route to galactose, a sugar that, as noted, is essential.

Lactose Intolerance

Adults, whether lactose tolerant or not, produce this sugar by steric rearrangement (epimerization) of one of the carbon atoms of the glucose derivative: UDP-glucose ↔ UDP-galactose. This "epimerase" is abundant in adult tissues but lacking in infants—not coincidentally, one may suppose. To repeat, as the source of galactose the infant relies solely on the lactose provided by the mother's milk and converts it to galactose with the aid of lactase. Once weaned from the mother's breast, and only then, the hitherto silent gene for the epimerase reaction pathway is turned on, assuring a permanent supply of galactose. One may argue—and admire Nature's wisdom—that for this reason, among others, the mother's breast becomes nutritionally dispensable during development and hence milk intolerance is a benign disorder.

Another potential mechanism for the age-related decline of lactase activity and hence increase in milk intolerance has recently been proposed. Rat intestinal lactase exists in two macromolecular forms: the active enzyme, a large trimeric molecule of 300,000 molecular weight, and an inactive monomer, one-third the size of the molecule formed by a proteolytic enzyme. After weaning, the gene responsible for this "processing" enzyme is turned on, which accounts for lactose intolerance. It is an attractive mechanism, though not yet shown to be responsible for the human disorder (Quan et al., 1990).

An exquisite control mechanism has evolved for the timing of lactose synthesis. In late pregnancy, as parturition approaches, the anterior pituitary gland of the brain sends a hormonal signal called prolactin to the breast. At the target site, prolactin triggers the formation of a "coprotein" known as α-lactalbumin.* The coprotein has the unusual property of modifying lactose synthetase; it is not a conventional catalyst. In the absence of coprotein, that is, prior to and again after lactation ceases, lactose synthetase catalyzes the formation of the lactose derivative galactosyl-N-acetyl glucosamine, a donor of sugar residues for glycoproteins found commonly on cell surfaces. The presence of coprotein during lactation suppresses formation not of lactose itself, but of this lactose deriv-

---

*The coprotein α-lactalbumin is one of a class of proteins that are not enzymes, and therefore not catalysts. These proteins change the specificity of an enzyme but do not accelerate reaction rates.

ative. The lactose synthetase becomes a catalyst for production of the milk sugar lactose, the source of galactose in the nursing infant. The reader will surely agree that the signaling process for milk production is beautifully timed and coordinated at the physiological and biochemical level.

As mentioned, neither lactose synthetase nor the coprotein α-lactalbumin is present in the sea lion's lactating mammary gland or in the milk this pinniped produces. This "abnormality" has been observed in all pinnipeds—seals and walruses as well as sea lions—characterized by webbed feet or fins, but curiously only in those inhabiting the Pacific Basin. My guess is that their nursing young "prematurely" develop the "adult" hepatic galactose-producing epimerase reaction (UDP-glucose → UDP-galactose). The lactose-lactase system can therefore not be said to be a universal evolutionary landmark of the mammalian species. Pinniped milk is unusually high in fat, nearly 40 percent, compared to less than 5 percent in human, cow, camel, and horse milk. In general, the higher the milk fat content of a given species, the lower the lactose content. There is no known metabolic explanation for this reciprocal relationship.*

The preceding discussion has attempted to rationalize the unique mammalian machinery for meeting a nutritional requirement of the young. Does lactose and lactase production with the goal of furnishing galactose occur at all in nonmammalian species, for example in birds and reptiles? The answer is no. As early as 1906, Plimmer stated that no animals below mammals, neither fish nor fowl, contain intestinal lactase. More likely, one suspects, they may rely on the "adult" mechanism, the glucose epimerase reaction, which one would suppose to be the "older" pathway for producing galactose.

The evolutionary significance of the mammalian lactose-lactase system remains puzzling. Lactose per se has no known function except as a source of galactose. The disaccharide occurs neither in microbes and plants nor in pinnipeds. Moreover, the universal adult mechanism for producing galactose by glucose epimerization is, at least superficially, less complex; only a single enzyme, the UDP-glucose → UDP-galactose epimerase, is needed. By contrast, the formation of the metabolically active form of

*One wonders whether prolactin occurs in the brain of pinnipeds, and if so, what its function might be.

galactose in nursing mammals begins with lactose synthetase, followed by lactase, the activation by ATP to galactokinase to form galactose-1-phosphate. Finally, a transfer of the uridyl group from UDP-glucose yields UDP-galactose. A total of four catalytic proteins is required. In addition, the enzyme lactose synthetase needs to be modified by the coprotein α-lactalbumin, the synthesis of which depends on the protein brain hormone prolactin. In sum, six proteins instead of one must intervene to produce galactose for the needs of the nursing infant—hardly an example of Nature's parsimony.

Lactase Deficiency

In humans the ability to digest lactose probably arose in prehistoric time, the Neolithic era, about ten thousand years ago, conferring a selective advantage on populations that practiced dairying (cattle, goats, sheep). Today a majority of the world's adult population (estimated at 70 percent), along with other mammals, seem to be milk intolerant. Whether the loss of lactase is an irreversible event remains an open question. In the contrary case one speaks of lactase "retention" or "persistence." Is this notion correct, or should adult lactase be referred to as "acquired," the enzyme induced by uninterrupted milk consumption?

Enzyme induction is a well-known phenomenon in the microbial world. The bacterium *Escherichia coli* adapts to growth on a medium containing lactose instead of glucose as a carbon source. Lactase, or β-galactosidase, as microbiologists call the bacterial enzyme for cleaving lactose to glucose and galactose, is readily induced.*

---

*To biochemists the term lactase is much less familiar than β-galactosidase. True, the latter refers to the mode of action, the catalytic act, rather than to the substrate on which it normally acts. The understandable reason for this terminology is that β-galactosidase became widely known as the classic example of an induced enzyme. The bacterium *E. coli,* when offered lactose in lieu of glucose, will use it as a carbon source, that is, it is induced to produce the cleavage enzyme β-galactosidase, which is functionally identical to lactase. Enzyme induction played a critical role in the formulation of the "operon

Animal experiments to resolve the issue of lactase persistence versus induction in adolescents and adults have given conflicting results. When fed lactose, young pigs whose lactase levels are falling develop symptoms typical of human lactose or milk intolerance and ultimately die. In similarly treated canine pups, lactase levels also fall initially; but when they are given high lactose loads, the deficiency symptoms disappear after a while, suggesting (but not proving) that the canine has resumed lactase production. Enzyme induction by a natural nutrient appears to be restricted to microbes. While the phenomenon of adaptation per se is well established in higher animals and humans, these higher organisms apparently respond only to substances foreign to the body, such as alcohol, drugs, and other molecules potentially toxic to the animal body.

In any event, the scientifically intriguing question remains of whether the infant enzyme and the adult lactase of lactose-tolerant humans are identical or distinguishable entities. Human experiments are feasible, but probably and perhaps properly forbidden on ethical grounds. A recent authoritative review on the subject states categorically, "Whatever the mechanism of the decline or of the persistence of lactase in adulthood, all attempts of reintroducing enzyme activity, 'adult induction,' have been singularly unsuccessful" (Semenza and Aurichio, 1989). Still, one wonders whether such experiments have been or can be properly designed. After all, when β-galactosidase is induced in *E. coli,* the bacterium responds to the substitution for one nutritionally essential carbon source (glucose) by another (lactose). In animals as well as man, such limiting conditions are not likely to prevail, especially when either glucose itself or glycogenic amino acids are available sources of carbohydrate (energy production) and as long as the epimerase reaction supplies galactose in adolescents and adults.

Up to this point I have reviewed current knowledge on the causes of milk intolerance—when, why, and where it occurs—in other words, the epidemiology of this benign affliction. What follows will be a look into the

---

hypothesis" by François Jacob and Jacques Monod (1961), pioneers of molecular biology. Their concept entails a set of intimately associated or contiguous genes whose expression is regulated by the same control elements, known as repressors and operators.

past at the uses of milk other than providing an essential infant nutrient. Once again, experiences from my youth have led me to dwell on this topic.

Food Preservation

Practices for food preservation must have played an important role in the survival and the prospering of early societies. Whether they were inventions or accidental discoveries, and how they came about, we will probably never know. They are lost in antiquity, transmitted from one generation to the next. Leavened bread, the making of cheese and wine, mentioned already in the Old Testament and known to the ancient Greeks and Romans, are fermentations—caused originally, one may presume, by airborne microbes, yeasts, or bacteria. No wonder that these food-preserving practices were discovered independently by societies widely separated geographically.

In central Europe, one such practice has survived to this day, or at least to the days of my childhood. *Saure Milch* was a fairly common item on my family's menu. A dish of milk was left on a windowsill in the summertime. After a few days in the open air, a yellowish semisolid skin, or layer, covered the surface. When sweetened, sour milk made an excellent meal—or so we were told by our parents. Obviously the milk did not simply spoil, as it does in the refrigerator once the container is opened. The explanation is that airborne bacteria, *Lactobacillus* species, thrive on milk, fermenting the milk sugar lactose to lactic acid,* and souring (sterilizing) the liquid. Once these bacteria have taken hold in their favorite medium, they grow vigorously and leave other airborne microbes no chance to compete.

Yogurt, kefir, and koumiss are all variations on the same theme: lactic

---

*Fermentation is the general term for the anaerobic dissimilation of sugars to smaller molecules. Fermenting yeast converts glucose to ethanol and carbon dioxide. Lactose fermentation produces lactic acid, $CH_3CHOH\ COOH$, by way of pyruvate, $CH_3\ CO\ COOH$, followed by reduction of the CO group to $-CHOH-$.

acid formation that removes lactose, souring and sterilizing the milk. Of Turkish origin, but now popular all over the Western world, is yogurt—milk fermented by *Lactobacillus bulgaricus*. Caucasian kefir is made much the same way, except that the starter culture—stock held over from a previously fermented batch—also contains *Streptococcus lactis* and yeast which convert some of the glucose to ethyl alcohol and carbon dioxide, resulting in an effervescent and slightly intoxicating brew. Finally, the nomadic Tatars, onetime inhabitants of Siberia and Mongolia, use the milk of their beasts of burden, mares and camels, as the source of koumiss, said to serve medicinal purposes as well. Modern microbiology provides the various milk-fermenting microorganisms in "pure culture," commercially available in powdered form, along with reproducible recipes. One need no longer rely on stock from a previous batch.

Milk Substitutes

Since World War II, immigrants from the Far East have become a growing sector of the North American population. Adult milk intolerance varies widely with ethnic background. Probably for this reason lactose-free milk is now available in food stores. It is made by treating the milk with the enzyme lactase, under the trade name Lactaid. Another product, known as "coffee creamer" and also lactose free, is based entirely on plant products such as soybean protein. Moreover, coffee creamers are entirely free of cholesterol, an added bonus for those who wish to reduce their intake of animal fat. The food industry has certainly risen to these challenges and incidentally has done so without protest from the adherents of "health foods."

Cheese Making

A variant of preserving the food value of otherwise perishable milk is the equally and probably more widespread practice of cheese making. It begins with the "curdling" of milk, the curd being a coagulum consisting largely of casein, (the principal milk protein), milk fat, and calcium. Curd, a semisolid mass, is produced by rennet, or rennin, an enzyme from calf stomach and probably other ruminants. Tradition credits this discovery to some ancient herdsman who, while tending his flock, observed the curdling of milk he carried in a bag made of leathery calf stomach. Curdling separates the water-insoluble milk components from the whey, the part that is watery and presumably contains much if not most of the milk's lactose. For this reason, one may speculate that either by design or by accident, cheese making not only provided a less perishable nutritious dairy product, but rendered it innocuous to the lactose intolerant. One wonders how many milk-intolerants are never recognized or diagnosed as such, nor are aware of their intolerance because the major dairy products they consume are butter and cheese. Experience probably has taught them to limit whole-milk consumption to a tolerable level, one or two glasses per day.

Residual lactose in cheese varies widely depending on the degree of desiccation of the curd. The moist, softest varieties (cottage cheese and mozzarella) retain some lactose, whereas the aged and hardest cheeses (Swiss, cheddar) are essentially lactose free. This advantage for persons intolerant of milk is accompanied by a relatively high content of choles-terol and saturated fat in the harder cheeses. Moderate consumption would appear to be the best nutritional advice generally, until food technologies succeed in manufacturing our favorites without impairing their palatability.

Global Distribution of Milk Intolerance

Data on the frequency of the lactase phenotypes (lactose persistence versus lactose intolerance) are not entirely reliable because they are not always based on the same diagnostic tests. Still, certain striking patterns in the global distribution have emerged (see Table 9.1). Lactose tolerance (or lactase persistence through adulthood) is predominant in Caucasian North Americans and in central Europeans and Scandinavians, ranging from 80 to 100 percent and nearly as high (about 75 percent) for the pastoral milk-consuming nomads (fermented milk, koumiss, or kefir?) of the Bedouin and Arabic tribes of North Africa. Intermediate levels (about 50 percent) are found in regions bordering the Mediterranean (France,

Italy, Yugoslavia, the Near East, Greece). By contrast, lactase deficiency is highest in India, Japan,* Taiwan, nearly all of mainland China, and among the Indian populations of Mexico, South America, but also in Native Americans of the United States and among Eskimos. A point of special interest is the relatively high frequency of lactose intolerance among Lapps (41 percent), which are tribes of Mongolian origin, compared with those of their neighbors, southern Scandinavian populations where lactose intolerance is exceedingly rare. Notable and so far unexplained are a few pockets of milk-tolerant populations in central Africa, the Hima and Tussi tribes in Uganda, the Fulani and Fulbi in Nigeria, and the Peuhl (Fulbi) in Senegal, while their immediate neighbors are among the least milk tolerant (Kretschmer, 1972). A history of pastoral cattle herding versus a nomadic lifestyle is a possible explanation.

Coda

A recent documentary film vividly describes the lifestyle of elderly men and women in the Georgian Caucasus. It seemed to this viewer idyllic—rather more so than that of senior citizens living in retirement homes. Travelers to the Caucasus report that the elder Georgians lead an unusually active life for their age, mostly outdoors tending their vineyards—which produce some of the best grapes and wines (Georgia No. 5)** in the former Soviet Union. A physician friend visiting the area was told that these elderly, an unusual number of centenarians among them, attribute their excellent health to regular consumption of fermented milk. Local

*Shortly after the end of World War II, the American government undertook massive relief programs to aid the undernourished populaces of India and Japan. The CARE packages contained various foods, among them powdered milk. Many of the recipients became ill, no doubt because they were milk intolerant. Rumors that the Central Intelligence Agency had a hand in the sinister design appeared in contemporary news stories.

**At the time I visited the Soviet Union in 1959, this was the label printed on the bottle. My hosts invariably recommended it as the best available in their country.

medical workers in Georgia also relate that the incidence of lactase deficiency is high* (statistical data for the region are not available). Still, the Georgians' health and longevity may have reasons other than the yogurt or kefir habit, which is not necessarily nutritional. The documentary makes the point that in Georgia the elderly are treated with great respect, indeed reverence, by the younger generation. They continue and actively participate in all communal activities.

A report in *Chemical and Engineering News* describes some behavioral studies relating social environment to the life span of rats. The animals, all elderly, were divided into four groups. In group 1, the control group, the rats were kept in their individual cages at all times. Group 2 rats were allowed once a day to leave their cages and run around. Group 3 rats were also taken out of confinement temporarily, but collectively, so that they could socialize with their contemporaries. Finally, group 4 rats were given the same temporary freedom as those of group 3, but their playmates were young rats. The results were predictable, but it is gratifying nevertheless that the rats of group 4 had the highest life expectancy.

The reader may wonder whether the concluding paragraph above is relevant to a chapter on milk intolerance. I present it solely for the reason that the conclusions one may draw from the lifestyle of the Georgian elderly or the socializing of rats would be out of place in any of the other chapters.

Bibliography

1. F. Roehmann and F. Lappe (1895), Ueber die Lactose des Duenndarms, *Ber. Deut. Chem. Ges.* **28**, 2506.

2. R. Quan et al. (1990), Intestinal lactose, shift in intracellular processing to altered, inactive species in the adult rat, *J. Biol. Chem.* **265**, 15822.

3. G. Semenza and S. Aurichio (1989), Small intestinal disaccharidases, in *The metabolic basis of inherited disease,* 6th ed., p. 2975. New York, McGraw-Hill.

*Alexander Leaf, Masschusetts General Hospital, Boston, private communication.

4. N. Kretschmer (1972), Lactose and lactase, *Sci. Am.* **227,** 70.

5. G. Platz (1989), The genetic polymorphism of intestinal lactase activity in adult humans, in *The metabolic basis of inherited disease,* 6th ed., p. 2999. New York, McGraw-Hill.

6. W. F. Enos et al. (1955), Pathogenesis of coronary disease in American soldiers killed in Korea, *J. Am. Med. Assoc.* **158,** 912.

7. F. F. McNamara et al. (1971), Coronary arteric diseases in combat casualties in Vietnam, *J. Am. Med. Assoc.* **216,** 185.

8, R. O. Castillo et al. (1990), Intestinal lactase in the neonatal rat, *J. Biol. Chem.* **265,** 15889.

9. D. W. Cramer (1989), Lactase persistence and milk consumption as determinants of ovarian cancer risk, *Am. J. Epidemiol.* **130.**

10. R. H. Plimmer (1906), On the presence of lactase in the intestines of animals: Adaptation of the intestine to lactose, *J. Physiol. (Lond.)* **35,** 81.

11. N. S. Scrimshaw and E. B. Murray (1988), The acceptability of milk in populations with a high prevalence of lactose intolerance, *Am. J. Clin. Nutr.* **48,** 1083.

10

Two Centuries of Pellagra Research

———

Herewith a historic review of the regional occurrence of

pellagra and practices for avoidance of this metabolic

disorder; also, an early example of definitive

epidemiology and the discovery of

vitamin deficiencies.

Pellagra ceased to be a public health problem after the discovery of nicotinic acid (niacin) in 1937 by Conrad Elvejhem at the University of Wisconsin. I knew little about this nutritional disease but became interested during the summer of 1957 when I visited the Mexican town of Oaxaca, famous for the excavations and remains of Mayan-related cultures.* It was market day, the central square in town bustling with activity. While my spouse was eyeing the beautiful Mexican rugs displayed by circulating vendors, my attention was caught by some elderly Indian women sitting on the sidewalks, selling chips of grayish-white stone piled in front of them. Not knowing Spanish, I was unable to extract any information about the nature of the chips and their uses, except that they

*The artifacts on view were those of the Mixtecs and Zapotecs, Mexican Indians who borrowed much of their culture from the Mayans.

had something to do with the preparation of tortillas. Colleagues I consulted later at the university in Mexico City explained that the stone was limestone, traditionally added to the grinding bowl when Mexican Indians prepared the cornmeal for tortillas, their principal source of nourishment.

I did not pursue this bit of information further until years later when I came across a brief paragraph in the chapter on vitamins in a biochemistry text by White, Handler, and Smith, (4th ed., New York, John Wiley, 1968), describing the deficiency disease pellagra and its cure and prevention by the "vitamin"* niacin, or nicotinic acid.** The same textbook passage mentioned the widespread occurrence of pellagra earlier in the century among the poor in the American South, typically in regions where corn—in this case, commercially milled cornmeal—along with pork belly were the dietary staples and the sole source of protein. The chapter concluded with an intriguing statement that pellagra was unknown among Indians in Mexico and elsewhere in Central America. Though poor, they subsisted on the same corn diets that appeared to be associated with pellagra in the southern United States. Also mentioned was what I had observed during my visit to Oaxaca, the Mexican custom of grinding corn in the presence of limestone and, most important, the hypothesis that the alkaline milieu created by a slurry of limestone might liberate nicotinic acid from some unknown precursor in the corn. Later I learned that the poverty-stricken Hopi Indians in Arizona and New Mexico, tribes who also subsisted on corn as the sole source of protein, never suffered from pellagra even though their diet was equally deficient in nicotinic acid per se.

*The quotation marks signify that nicotinic acid is not a vitamin in the strict sense. It can be synthesized by most animal species.

**Nicotinic acid is known to the general public only under the trivial name niacin. Apparently "nicotinic acid" is unacceptable for two reasons. The designation "acid" is odious because acids are bad for you (stomach acids, for instance). For similar reasons ascorbic acid, the antiscurvy or antiscorbutic vitamin, is sold under the name vitamin C. Moreover, "nicotinic acid" is taboo because it implies, and in fact does have, a chemical relation to nicotine. It was first made by chemical oxidation of nicotine, a prominent alkaloid of tobacco.

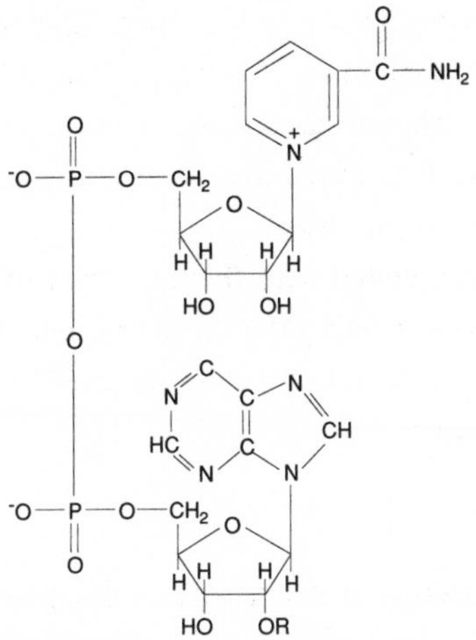

Fig. 10.1 Structure of NAD, nicotinamide-adenine dinucleotide, a coenzyme functioning as a hydrogen acceptor and, in its reduced form NADH, as a hydrogen donor.

The chemistry underlying the wisdom of Mexican and Hopi Indian practices remained a mystery until quite recently. Mature corn is inadequate as the sole source of nicotinic acid because zein, its principal protein (unlike that of other cereal grains) is deficient in tryptophan, an essential amino acid. In turn, tryptophan is normally a precursor or source of niacin in the animal body. What then is the material in corn that furnishes niacin on treatment with the alkaline limestone? The latest evidence suggests that it is mainly NAD (nicotinamide-adenine dinucleotide; Fig. 10.1), a complex coenzyme* that occurs abundantly in corn. But NAD in humans, unlike niacin or niacin amide, is not absorbed efficiently from the gastrointestinal tract into the bloodstream and therefore is nutritionally unavailable to man. The NAD is indeed sensitive to alkali and, under conditions that simulate the Mexican practice, releases niacin.

While this biochemical mystery is largely solved, others remain that have nothing to do with the sciences of biochemistry and nutrition. Why, I wondered, did knowledge of the Mexican Indian (or Hopi) practices

---

*Coenzymes have no catalytic properties per se. They function only when combined with an enzyme protein, sometimes called an apoenzyme.

never cross the "green border" into the North American continent, remaining unknown for so long in the American South? Was it xenophobia? Perhaps anthropologists should have paid attention; but in the early days of the century their interests may have been elsewhere, not in mundane matters such as nutritional practices.

My account could have ended with this summary of the essential facts. Yet the more I read about pellagra the more fascinated I became with the history of the disease. What follows may be unnecessarily detailed and test the reader's patience.

Foundation of the Science of Nutrition

By the middle of the twentieth century the science of nutrition had reached its primary goal, at least as perceived by its practitioners at the time. Man and numerous animal species could be raised and maintained on fully synthetic diets, regimens that contained exclusively chemically identified components. The prerequisite for this achievement was to isolate, purify, and establish the structure of those molecules that animals are unable to synthesize on their own: by definition, the essential amino acids, essential fatty acids, and vitamins. In many cases these growth factors were originally detected in plants, microbes, or yeast and, when absent in food, shown to cause dietary deficiency diseases. Much of the science of nutrition was in fact a multidisciplinary effort, requiring the combined skills of physiologists, organic chemists, and biochemists.

The principal task of the nutritionist is no longer the identification of essential or growth factors. In fact, none have been found for several decades. The nutritionist's aim today is to determine the quantities of the various dietary components needed for optimal growth and maintenance of health. This endeavor is much more difficult and often controversial—confusing to the public—for the reason that some of the requirements depend on age, body weight, gender, lifestyle, and so on. Federal agencies here and in other countries might address this problem by setting ranges rather than target single numbers, especially for vitamin and mineral requirements. Yet the chosen terminology "RDA," or recommended daily allowance, reflects one of the dilemmas facing public health policies; in my view, it is a compromise. The term "allowance" can be taken to mean that a little more or a little less will not hurt or, for that matter, benefit the individual. The RDAs are still periodically revised, but in recent years the numbers have not changed significantly. Moreover, it seems a waste of effort and funds for several Western countries each to issue their own RDAs. In any event, they do not differ substantially.

An important objective of ongoing nutritional studies is to elucidate the specific biochemical functions that essential growth factors serve. This endeavor is complicated by the fact that several of the vitamins (C and perhaps E and K are possible exceptions) and other essential molecules exert their biological effects only after conversion into other molecules, the majority of much greater complexity. This subject, known as intermediary metabolism, remains a major focus of biochemical research and its methodology. Anyone who has looked at a map of metabolic pathways summarizing half a century's research will be impressed by the intricate patterns, perhaps comparable in complexity to what a London taxi driver needs to know in order to be licensed. Still, some practical benefits accrue from this knowledge, which is best acquired by memorization and hardly ever by logic. In a growing number of cases the unraveling of intermediary metabolism has provided the clues for designing antimetabolites, foreign compounds that have specific therapeutically valuable effects.

The classic and earliest examples of vitamin deficiencies were discovered in individuals exposed to extreme dietary deprivation, not general malnutrition (lack of calories). A few of these familiar stories are retold here, only because they have certain elements in common with the pellagra story, which is less widely known.

Some credit must go to the shipboard physician James Lind for performing the first controlled nutritional experiment in the 1750s. The enfeebled sailors under his care on a long sea voyage he divided into two groups. One received supplements of fresh vegetables; the second, the untreated group, served as controls. Without exception every seaman in the experimental group, but none of the controls, recovered quickly from the deficiency disease. Lind had discovered a cure for scurvy, eventually shown to result from the lack of ascorbic acid. It was a very clear-cut experiment, requiring no statistical evaluation or confirmation.* The fresh fruit or vegetable remedy for scurvy was soon adopted by His Majesty's Navy, hence the nickname "limeys" for British sailors.

Similarly straightforward was the discovery of beriberi, a widespread nutritional deficiency disease in East Asian countries, where rice is the predominant staple. The content of the B_1 vitamin thiamin is naturally low in rice, and it is almost totally absent in the "refined," or polished, grain. Wild rice grown primarily by Native Americans is unrefined and therefore, one may presume, nutritionally adequate. It may be one reason why beriberi has been an unknown disease on the American continent. In this instance, agricultural technology has not been an advance—quite the contrary. Modern man's craving for foods that are refined, preferably colorless, and with an extended shelf life, has certainly proved of dubious benefit. Processed staple foods are now routinely "fortified" and enriched with a number of vitamins. We are putting back what has been artificially removed—as mandated by government regulations in Western countries. The jury is still out on the ultimate effects on both health and economics.

Like scurvy, beriberi, and rickets,** cases of pellagra are rarely seen any longer. Few practicing physicians encounter a patient afflicted with one of these nutritional diseases. By 1947 pellagra had been eradicated. Still, isolated cases continue to appear sporadically and some physicians have expressed concern. To quote J. L. Spivack and D. L. Jackson (*Johns*

---

*According to Kenneth J. Carpenter, *The History of Scurvy and Vitamin C* (Cambridge, Cambridge University Press, 1986), greater credit must go to Captain Cook in the eighteenth century. The real change that reduced scurvy in the British navy did not come until the Napoleonic Wars, when citrus juice was routinely added to sailors' food (J. T. Edsall, private communication).

**A rickets epidemic in the English Midlands and Scotland was reported some years ago, but the diagnosis was slow in coming. Few if any of the younger physicians had ever seen a case. The epidemic was localized in communities settled by recent Asian immigrants from Pakistan and India, especially in children and adolescents. The lack of sunshine in their new homeland, their low intake of milk, and their clothing habits combined to make them deficient in the "sunshine" vitamin, D (editorial on Asian rickets in Britain, *Lancet,* August 1981).

Hopkins Med. J. **140** [1977], 295): "Since many physicians have never seen a pellagrin, these patients can present a diagnostic problem. We consider it important for physicians to be familiar with the clinical manifestations of the disease." Their paper describes in detail the symptoms of fourteen pellagrins among fifteen thousand patients admitted over a ten-year period to the Johns Hopkins Hospital. The authors obviously felt that medical students must continue to be made aware of any disease, no matter how rare. One may add that such knowledge, even if never or seldom applied in practice, should be part of the physician's "liberal" education.

Early Pellagra Research in Europe

Spanish, Italian, and French physicians were in the forefront of recognizing pellagra, the disease of the four Ds: dementia, diarrhea, dermatitis, and ultimately death. It is a detective story of the kind that in modern times Berton Roueché has recorded so masterfully in his "Annals of Medicine" for the *New Yorker.* But what makes the history of pellagra unique is that it extends over a period of two centuries. Christopher Columbus introduced corn (Indian maize) to the European continent, and only gradually did this novel crop become popular as a dietary staple. Thus, there are no early records anywhere prior to the eighteenth century of pellagra as a corn-related nutritional disease. Yet this scourge is estimated to have caused half a million deaths worldwide between 1730 and 1930. Historians credit the Spaniard Casals, physician to Philip V, with the earliest description (1732) of an ailment common among Asturian peasants. He named it *mal de la rosa,* characterized by erythema or redness of the skin. Apparently his paper was not brought to the attention of the medical profession in the United States prior to a publication of Thiery in 1932.

Early in the nineteenth century, Italian physicians named the disease pellagra (from *pelle agra,* rough skin). They were the first to discuss pellagra's epidemiology and etiology with remarkable insight, even by modern standards. Their experience with malaria, the disease rampant in southern Italy, trained them well for asking the right questions. Buniva, practicing in Turin in 1805, described the disease as one "contaminating the beautiful cisalpine and subalpine countryside, one that attacks the *poor* peasants almost exclusively." He concluded, "Nothing gave me any reason to suspect that this disease can be communicated."

Equally remarkable was the proposal of Filippo Lussana and Carlo Frua in Milan (1856) that ascribed the disease to nutritional imbalance: "It is our hypothesis that pellagra originates and proliferates whenever the diet lacks protein (nitrogenous substance)." In essence, the Italian physicians had recognized the etiology of pellagra.

One wonders why these early views remained unknown to the U.S. medical community. Even Joseph Goldberger's classic pellagra studies, certainly the early ones during the 1920s, failed to refer to the Italian hypothesis. Perhaps lack of communication is to blame. Today we take instant, worldwide information transfer for granted. But until the nineteenth century, journals comparable to *Lancet* (founded in 1832) or the *New England Journal of Medicine* (founded in 1812) did not exist, nor did international meetings for sharing information. Language barriers may have been another obstacle. Only after World War II did English become the *lingua franca* of medicine and science. Even if the pellagra literature had been known on the American continent, it might have been disregarded because the views of the Europeans were by no means unanimous. And perhaps most important, pellagra was unknown in the North American continent until this century.

Relevant to the above discussion is a French doctoral dissertation of 1863 by I. Salas on the "Etiology and Prophylaxis of Pellagra." He reached the conclusion that the cause of pellagra is not corn per se, but "spoiled" corn, infested with a parasitic endophyte called verdet. Salas also made the strong point that in the southern United States, with a population of 20 million people and corn commonly used as a staple, not a single case of pellagra was on record in the nineteenth century. Similarly, in Mexico the disease was unknown. Salas did mention the Mexican practice of cooking the maize with limewater. This practice, Salas said, "will stop the growth of the parasitic fungus (verdet)." His countrymen in Burgundy were using the same Mexican practice, assuming that it prevented the corn from spoiling. One could not and cannot take exception to Salas' arguments, yet they led him to the wrong conclusion.

The "spoiled corn" theory had its adherents elsewhere, as reflected by the contemporary admonition of Italian health authorities to "buy wheat bread, use little polenta, and include some milk and cheese in the food you eat." As late as 1910 L. W. Sambon, an investigator at the renowned London School of Hygiene, used much the same argument, attributing the cause of pellagra to "unsound maize" (*J. Trop. Med.* **13,** 289), naming blood-sucking insects, black flies, and buffalo gnats as the disease vectors.

If American medical and public health authorities paid little attention to pellagra, or perhaps were even unaware of the disease, who was to blame? Perhaps historians have the answer. To me, the simplest explanation seems to lie in the apparently total absence of the disease until the early 1900s. At that time, the U.S. Public Health Service was preoccupied with the eradication of yellow fever and typhoid fever, then endemic in the South.

Pellagra was apparently first noted in the United States around the turn of this century, and its symptoms were recognized to be typical of the same poverty disease earlier prevalent on the European continent. In 1907 a total of eighty-eight cases was reported from an insane asylum in Alabama. Eventually, in 1914, the Public Health Service became sufficiently alarmed to appoint Joseph Goldberger,* one of its ablest staff members, to take charge of what threatened to become a scourge of epidemic proportions.**

I interrupt here to recount a seemingly unrelated but historically important discovery of an animal deficiency disease known as "black tongue," or typhus, in dogs. This canine disease was characterized at the turn of the century by a German veterinarian named Hofer. A decade later, in 1916, T. N. Spencer, a veterinarian practicing in North Carolina, raised the question, "Is black tongue in dogs pellagra?" His answer was affirmative. He was able to cure black tongue by feeding dogs protein-

---

* Nearly all of Goldberger's work appeared in the form of reports to the U.S. Public Health Service. One wonders how widely they were read. I owe reprints to the late DeWitt Stetten of the National Institutes of Health.

**Arthur Kornberg, in *For the Love of Enzymes* (Cambridge, Massachusetts, Harvard University Press, 1989) pays admiring tribute to Goldberger, one of his heroes.

rich meals—meat, milk, and eggs, a regime that had been successful in treating some human pellagra cases.

Identical symptoms of the canine and human diseases were observed. Proof that dietary deficiencies were the cause came in the 1920s, notably from the researches of Goldberger's group. Dogs, the only animal model for many years, proved critical for the final solution to the problem—today's opponents of animal experimentation take note. Nutritional experiments are carried out and controlled much more easily with animals than with humans. Admittedly the choice of species is often arbitrary or accidental, rather than rational (see Chapter 11). Still, more than a decade would go by before Conrad Elvejhem and his collaborators at the University of Wisconsin were able to isolate the anti–black tongue factor, nicotinic acid, now given the name niacin. The reason why in the past some populations contracted pellagra and others, on seemingly identical diets, did not is and in part remains an intriguing problem of nutritional history.

The following section describes in some detail why Goldberger's systematic research and the canine black tongue connection were vital to the solution of the pellagra problem.

Chronology of Goldberger's Research

It is not obvious from the literature how much Goldberger knew about pellagra when he began planning for the task assigned to him by the U.S. Public Health Service in 1914. His first papers do not mention the earlier European literature. Still, the design of his initial approach leaves no doubt that he was aware of the controversy over whether pellagra was an infectious disease or one of nutritional deficiency. He was able to resolve this issue once and for all by studies which to me and others have remained models of epidemiological research. For a whole year (1914–1915) he visited insane asylums and prisons in several pellagra-ridden southern U.S. states, inquiring into sanitary and nutritional practices.

Goldberger's earliest but not yet conclusive evidence favoring a dietary origin of pellagra appears to be based on a study of eleven convicts. During the period of observation six of these Alabama prisoners came down with the disease. Why the ambiguous result? Here is one possible explanation.

According to the protocol, the eleven volunteers received coffee twice daily; how many cups they drank is not recorded. Yet roasted coffee is known to contain substantial amounts of nicotinic acid. This dietary variable seems to have received little attention in the pursuit of pellagra epidemiology, including those studies dealing with the puzzle of the pellagra-preventive practices among Mexican Indians. At any rate, the somewhat preliminary results of Goldberger's first venture were hailed in the press. The front page of the *Jackson* (Mississippi) *Daily News* (November 1, 1915) carried this headline: "Goldberger Produces Pellagra among Convicts."

Perhaps Goldberger's best-known and most definitive early work was the paper reporting on inmates in a Mississippi prison where pellagra was rampant (Goldberger and Wheeler, 1915). The control group, the attending medical staff, never contracted the disease. Upgrading the prisoners' diet to what the nurses and physicians received dramatically improved the convicts' condition, confirming what one faction of Italian physicians had much earlier postulated. The hypothesis that pellagra is a contagious disease was no longer tenable. Having recognized that pellagra typically was poverty related, Goldberger was quoted in the article as saying, "I am only a bum country doctor and what can I do about the economic conditions in the South?" What indeed could he do about the price of cotton, which was stated to be inversely related to the incidence of pellagra?

In his next study, reported in 1916 to the Academy of the Southern Medical Association, sixteen healthy volunteers—Goldberger was one of them—were injected with "infectious" materials from the pellegrin's blood, nasopharyngeal discharges, and scales from the skin. After four to six months none of the brave volunteers showed any of the typical pellagra symptoms. Still, Goldberger and his colleagues (1918) were cautious: "The hypothesis that pellagra is of dietary origin, not contagious, is greatly strengthened."

Focusing next on the nature of the essential nutritional factors, Goldberger undertook an epidemiological study of several thousand individuals from eight hundred families living in small villages of South Carolina. The results were clear-cut. Only on diets containing ample meat, eggs, and milk (rich in vitamins) did individuals remain in good health. By 1918 it was evident that Goldberger was on the right track. Pellagra was a nutritional disease caused by the lack of "good protein and vitamins" (Goldberger et al., 1918). He could not go much further. At the time, none of the vitamins were chemically identified—only their major sources, such as yeast and liver extracts. Nor was it known that some proteins, especially of cereals, were low in essential amino acids such as tryptophan. Yet Goldberger's conclusions were guarded, perhaps because Elmer V. McCollum, one of the most influential and distinguished nutritionists of the time, still held the view that pellagra was contagious, "that poor diet predisposes to infection as do unsanitary conditions" (*A History of Nutrition,* Boston, Houghton Mifflin, 1957).

The last phase of Goldberger's research, from 1922 to 1928, edged toward a complete solution of the pellagra problem but unfortunately did not achieve the ultimate goal. Still, two important papers appeared during this period.

It was in 1922 that Goldberger and Tanner first acknowledged and confirmed what the Italian physicians Lussana and Frua had postulated in 1856, that pellagra is due to "alimentation of proteinaceous insufficiency." They were able to correct what they suspected to be an amino acid deficiency with a mixture of tryptophan and cysteine. Finally, Goldberger's laboratory provided definitive evidence for the identity of experimental black tongue in dogs and human pellagra (Goldberger and Wheeler, *Pub. Health. Rep.* **43** [1928], 172). Since the animal model was valid, the stage was set for biochemists to isolate and chemically identify the antipellagra factor.

No major papers on pellagra were published for nearly a decade after Goldberger's death in 1928. It was no longer a public health problem. In the meanwhile, biochemical developments of outstanding importance took place during the search for cofactors for glucose fermentation and respiration. In 1934–1935 the laboratories of Otto Warburg in Berlin and of Hans v. Euler in Stockholm showed that a cofactor for the oxidation of glucose-6-phosphate to 6-phosphogluconic acid, called cozymase, contained inter alia one unit of niacinamide (see Fig. 10.1). These were heroic experiments,* involving the fractionation of 100 liters of horse blood. (Warburg, an avid horseman, maintained a sizable riding stable.) Nicotinic acid as such had first been isolated from natural sources (rice bran) by Suzuki and coworkers as early as 1913. In fact, it has been available chemically by degradation of nicotine since 1848. The aim of the Japanese workers was to identify the antiberiberi factor, now known as vitamin B_1 or thiamine, an attempt that failed. At the time, no one tested niacin's effect on pellagra.

The scene shifts to Madison, Wisconsin. It is not clear exactly when Elvejhem's laboratory began to search for the factor that cured black tongue disease. In the early 1930s all experts apparently agreed that the symptoms of pellagra and canine black tongue were strikingly similar. Moreover, yeast and liver extracts were active in curing the deficiency disease in both humans and dogs—strong evidence, but not yet proof. All efforts to concentrate the active principle in pure form led instead to the isolation of "lactoflavin," now called riboflavin, a vitamin and precursor of another coenzyme involved in carbohydrate oxidation (respiration).

For several years the antipellagra factor eluded identification. Elvejhem's laboratory also had assay problems. Initially, they used chicks which showed pellagra-like symptoms on deficiency diets. Since the results with these animals were at best suggestive, the decision was to test the canine black tongue syndrome as a valid and reliable animal model.** The famous 1937 paper of Elvejhem and his colleagues reported two definitive findings in the following order: a single dose of commercial (Eastman Kodak) niacin gave a phenomenal result with a single dog suf-

---

*As a newly arrived graduate student in the Department of Biochemistry at the College of Physicians and Surgeons in New York I heard the electrifying announcement of Warburg's research, the subject of a departmental seminar. We all sensed that his work was the most important advance in biochemistry since the days of Pasteur. Moreover, Warburg's discovery of nicotinamide as a component of cozymase became essential for the later identification of the antipellagran factor by Elvejhem.

**Probably this was a reluctant decision. Purebred chickens were readily available, whereas dogs from the pound were invariably mongrels.

fering from black tongue; and secondly, nicotinic acid amide had been isolated in pure form from highly active liver concentrates (supplied by Eli Lilly & Company). Whether one of these experiments preceded the other is not stated. Presumably they were done simultaneously and independently by members of the Wisconsin team.

In turn, the pure "liver factor" and the chemically produced nicotinic acid (or its amide) were equally effective in curing the canine black tongue disease. The Elvejhem paper concludes with a still-guarded statement: "That a deficiency of this material may be the cause of black tongue is most interesting. Clinical trials are needed to show the same effectiveness in human pellagra." Such experiments were carried out immediately by T. D. Spies and associates (1939).

Remaining Pellagra Problems

In defense of the "spoiled corn" hypothesis that had been a contender for so long, it can be said that its adherents had some intriguing points. Poverty may be a cause either of nutritionally deficient diets or else of food that is spoiled, not fresh. The poor are more likely to consume stale food infested by microbial contamination.*

A number of factors contribute to the epidemiology and the tangled history of pellagra—why and where it occurred. In the United States the disease typically erupted among institutionalized patients, but rarely in the countryside. Most likely the source of institutional cornmeal was commercial, mechanically milled and degermed, whereas corn harvested

---

*To cite another example of spoiled-food speculation, some historians suggest that the Salem witches were in fact the victims of ergotism, reduced to eating rye bread infested with the fungus *Claviceps purpurea*. This microbe produces the pharmacologically active ergotamine. When taken in large doses, the alkaloid is known to paralyze the sympathetic motor nerve endings and hence is responsible for muscular twitching (and perhaps hallucination?). The same accounts also report an effective cure for the bewitched. Under the care of the village priest the victims were given fresh rye bread to eat and recovered promptly. Needless to say, this remedy greatly increased belief in the priest's supernatural power.

Table 10.1

Nutritional availability of nicotinic acid (NA) from mature corn

Source	Free NA (gm/kg)	Weight gain of rats (gm)	
		− NA	+ NA
Raw (mature) corn	0.4	0.4	22.3
Boiled in water	3.8	6.8	21.9
Cooked in limewater	24.6	22.3	25.0

Source: Carpenter, 1980.

by individual farming families was homegrown and perhaps stone-ground. That only some members, not whole families, came down with pellagra strengthened the notion that the disease was not contagious. It was the mother and housewife who was most often afflicted. In fact, Goldberger made the point that 70 percent of all the pellagrins were women. The little meat the family could afford was reserved for the breadwinner, and the milk for the children. Only cornbread, the cheapest food, did the mother take for her own nourishment. We may assume that hominy grits, the staple for the poor of the American South, Central American Indians, and Native Americans, was homegrown field or Indian corn.

As for the nicotinic acid content so vital to pellagra prevention, there are important differences related to the ripening process and to treatment of the cornmeal. Immature corn, whether sweet (milky) corn or field corn (maize), harvested twenty days after pollination, generally contains more nutritionally available nicotinic acid than mature corn, harvested about eight weeks after pollination. Prolonged boiling of mature corn in water releases about 20 percent of the total niacin in the free or nutritionally available form. Finally, alkali treatment, 1 percent limewater at 80° C, liberates nearly all of the "vitamin," rendering it nutritionally available. Some of these results are shown in Table 10.1.

One can argue that the corn plant, while growing from the milky to the mature state, needs little free niacin or niacinamide but large amounts of the niacin-containing coenzymes NAD and NADP for energy production. In corn that is mature or ripe, these coenzymes are stored not as such but linked to macromolecules, probably starch. The latter, the bound forms of niacin, are water insoluble and nutritionally unavailable because they do not pass from the gastrointestinal tract into the bloodstream. These bound forms release niacin only on treatment of mature corn with alkali, that is, when cooked with limewater. Experiments were carried out eventually in Cambridge, England, by Carpenter, Kodicek, and Wilson (1960), in imitation of the Mexican Indian practice (Table 10.1).

We may also ask why pellagra developed *only* on diets containing mature corn or maize as the staple (70 percent or more). Diets consisting of other cereal grains, wheat, or rice do not cause the disease, perhaps but not necessarily because their content of free nicotinic acid, or more likely of the coenzymes NAD and NADP, is adequate.

There remains one aspect of the pellagra problem that I find intriguing, a procedural detail of ancient practices to prevent the disease by treating corn with lime. A 1947 account by Sylvanus G. Morley describes the procedure minutely. I quote only relevant portions here:

The daily work of preparing corn for tortillas was and still is the most important activity in the life of Mayan women. This domestic activity can be divided into several steps. (1) The dry threshed corn is placed first in a pot (*kumm* is Mayan for tank) to be cooked, with sufficient water and lime to soften the grains. The mixture is heated until it is about to boil and is kept at this temperature until the hull is soft, and once in a while it is stirred. Then the pot is put to the side and left to rest until the following day. The soft corn is called *kuum* in the Mayan language. (2) The following morning, shortly after breakfast, the *kuum* is *washed until it is perfectly clean* and free of hulls. (3) Next, the *kuum* is milled. . . . The ground corn, which is called *zacan* in Mayan, is covered with a napkin and left covered until later.*

After these preliminary operations, the Mayan woman is ready to bake the tortillas. According to the information I gathered in Oaxaca, the first step prior to preparing tortillas consists of grinding the dry corn in the presence of limestone. According to the quoted account, the dry fresh corn is first cooked, not ground, with sufficient water and lime to soften the grains. This may be a trivial point, but the following step in the Mayan

*Translated from the Spanish by Susan Bloch.

procedure—the washing of the softened corn until it is "perfectly clean"—creates a problem.* If lime treatment releases free nicotinic acid from that bound to NAD or NADP, or for that matter uncharacterized glucose polymers (starch?), washing the kernels should transfer the freely water-soluble "vitamin" to the wash water. This the Mayans apparently discarded. In fact, the cited laboratory experiments of Carpenter and Kodicek, by and large imitating the Mayan recipe, took great pains to avoid this potential loss. In their experiments the wash water left from the lime treatment was saved and recombined with the "clean" corn. Thus, while modern science confirms the wisdom of the lime treatment that releases bound nicotinic acid, the subsequent manipulations performed by the Mayans should have negated it. The pellagra-preventing principle should have been "lost in the wash." The puzzle remains unsolved.

The poverty-stricken Native Americans, subsisting on corn, escaped pellagra by a variant of the Mayan practices. For preparing their *piki,* as they called the equivalent of the Mexican tortillas, they treated maize with wood ash (potassium carbonate), which is also alkaline. This method was and perhaps still is practiced by the Hopi in Arizona. Other tribes had learned that immature "milky" maize, when baked, roasted, or simply boiled (without lime), provided them with a nutritious diet.** Neither they nor the Hopi contracted pellagra.

Tryptophan, the Normal Source of Niacin

Cereal proteins vary markedly in nutritional value, depending on their content of the various essential amino acids. One of those is tryptophan. The principal corn protein—and in turn the major if not the only protein

---

*During a recent visit with colleagues in Tallahassee, Florida, I heard (but lack any documentation) that in some instances African-Americans drank the "pot liquor" and therefore may have escaped pellagra. Was this also a practice among Mexican Indians?

**The origin of popcorn? A type of Indian corn (*Zea mays erecta*) which on exposure to dry heat popped, that is, was "everted by the explosion of the contained moisture" (Webster's unabridged, 2nd ed; 1951).

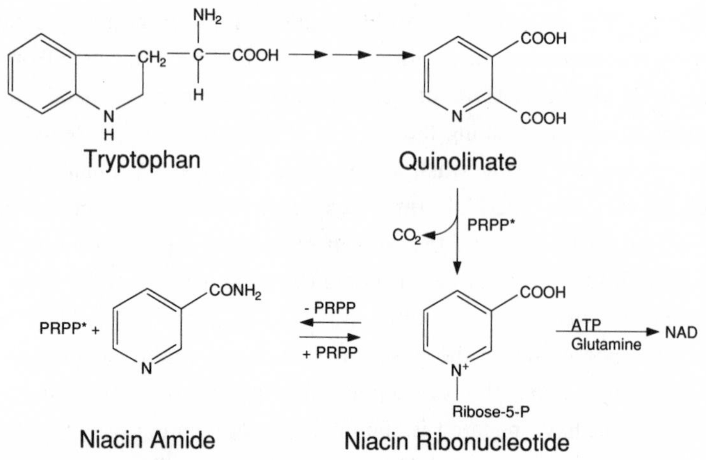

Tryptophan   Quinolinate

Niacin Amide   Niacin Ribonucleotide

* phosphoribosyl-pyrophosphate

Fig. 10.2 Conversion of the essential amino acid tryptophan to niacinamide.
The first three steps are catalyzed by oxygenases.

source in tortillas—contains minimal amounts of tryptophan—less than
0.05 gm per 100 gm of corn.

Probably tryptophan is an essential structural element of most proteins,
but more important for the pellagra story, it is the sole source of nicotinic
acid in unsupplemented diets. A complex system of seven enzymes discovered by Nishizuka and Hayaishi (1962) degrades the tryptophan structure to the "vitamin" instead of constructing it directly from smaller building stones. The abbreviated pathway is shown in Fig. 10.2.

Not all of the participating enzymes are necessarily present in various
animal species. Felines are a known exception. One of the enzymes converting tryptophan to nicotinic acid (niacin amide) is absent in domestic
cats and presumably other felines. All members of the cat family are
obligate carnivores, deriving essential nutrients from their prey (see Chapter 12). If investigators were to raise cats on fully synthetic diets, nicotinic
acid would have to be included. Such experiments have apparently not
been done, but it is known that several other enzymes are missing in cats,
which explains their predatory lifestyle.

In the context of pellagra, nicotinic acid is of course an essential dietary supplement—but for different reasons, as we have seen. Some authors dealing with pellagra classify nicotinic acid as a vitamin, others do not. At any rate, the current practice in food labeling for humans is to list nicotinic acid (niacin) in the vitamin category.

There is at least one reason why the pathway to nicotinic acid chosen by nature appears unnecessarily complex. Tryptophan, apart from nicotinic acid, produces serotonin, a neurotransmitter and vasoconstrictor, and also the plant growth hormone indoleacetic acid. With few exceptions, amino acids, whether essential or not, serve multiple functions; their role as structural elements in proteins is only one. If Tryptophan is indeed the sole normal source of nicotinic acid, supplementing corn diets with the amino acid should be as effective in treating pellagra as nicotinic acid. This has indeed been shown to be the case in both rats and human patients.

As we have seen, pellagra was conquered or prevented in earlier times by ancient practices that release bound nicotinic acid, or by shifting from corn to dietary proteins rich in tryptophan. Today's approach to the problem would undoubtedly employ genetic techniques. In the past, classic plant hybridization produced hundreds of strains or varieties, with the principal aim of obtaining characters such as improved yields per acre as well as disease and temperature resistance. This was the great achievement of what is known as the green revolution, pioneered by Norman Borlaug. As a result, many Asian countries have now become self-sufficient and are independent of cereal imports, especially rice and wheat.

As for corn, southern leaf blight in the United States has been eliminated. A variety known as maize-opaque-2, enriched in the essential amino acid lysine, is now available to farmers. One major effort of current plant biotechnology, still in the experimental stage, is to introduce into corn genes coding for biochemically well-defined characters. This genetic engineering promises to be the second phase of the green revolution and may well lead to tryptophan-enriched zein, the principal protein of corn. The alternative to genetic engineering is to fortify deficient diets with essential amino acids as such, but so far this approach has been too costly.

I   II   III   IV

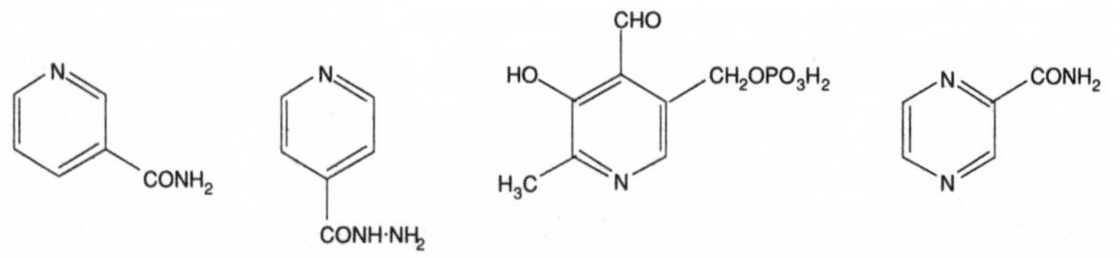

Nicotinamide   Isoniazid   Pyridoxal Phosphate   Pyrazinamide

Fig. 10.3 Structures of drugs (II and IV) based on the antimetabolite concept and their presumed coenzyme targets (I and III).

Addenda

It is a long-standing practice of pharmaceutical companies to screen synthetic compounds accumulated on their shelves for chemotherapeutic effects. This hit-or-miss approach paid huge dividends with the discovery of the sulfonamide prontosil (G. Domagk) and of DDT (P. H. Mueller) in the 1930s and 1940s. Once the structures of essential growth factors became known, a more rational approach, the antimetabolite concept, replaced what in the past had been essentially accidental discoveries. Drugs based on the antimetabolite concept were first designed to inhibit the growth of bacteria by interfering with a known metabolic function.* The targets were enzymes or coenzymes; the bacteria were deceived into accepting chemically designed analogs of the metabolite, rendering them nonfunctional.

As an example, niacin or nicotinamide as an antimetabolite target has been mentioned here. In 1952 the Hoffman–La Roche company received a patent for introducing isoniazid as a potential drug for the treatment of tuberculosis, while the earliest patent granted for this class of tuberculostatic analog, for pyrazinamide, was held by E. Merck and dated to 1936. The structures of these antimetabolites and their targets are shown in Fig. 10.3.

*Sulfonamides proved to be antimetabolites for *p*-aminobenzoic acid, a component of the vitamin folic acid, but the discovery was serendipitous. Folic acid is used inter alia in the biosynthesis of nucleic acids.

The structures of both II and III in the figure resemble nicotinamide (I) more than IV, which is a pyrazin containing two nitrogen atoms in the ring system rather than a nicotinamide derivative. Both drugs proved to be exceptionally potent bacteriostatic and bactericidal agents for *Mycobacterium tuberculosis*. Along with streptomycin they have been the effective weapons for the eradication of this age-old disease.

The intended mode of action of these antimetabolites is (or was) to displace nicotinamide (I), preventing the natural substrate from combining with the bacterial enzyme that forms NAD. Since the coenzyme pyridoxal phosphate (III), isolated by E. Snell in 1936, involved in amino acid metabolism is also a pyridine derivative, and since isoniazid inhibits pyridoxal phosphate-dependent enzyme reactions as well, the possibility exists that isoniazid, when inhibiting bacterial growth, has more than a single target.

As mentioned above, pyrazinamide became available as early as 1936, shortly after Warburg's characterization of NAD as a nicotinamide-containing coenzyme and simultaneously with Snell's discovery of pyridoxal phosphate (III), but a year before Elvejhem's discovery of niacin as the antipellagra factor. This chronology raises the possibility that pyrazinamide was actually designed as a pyridoxal antimetabolite and not as an analog of niacinamide. The rationale led to an effective antimetabolite, but the bacterial target may not be the one, or the only one, aimed at. I have been unable to clarify this issue.

Studies on the alarming growth of antibiotic resistance to the human tubercle bacillus have pointed to what appears to be the true action of isoniazid. The drug per se is not the antibiotic, but only when oxidatively modified in vivo by the enzyme catalase. Genetic catalase deficiency may be the key to antibiotic resistance (Douglas et al. [1992], quoted in *Science* **257**, 1038).

The fruits of pellagra research would ultimately benefit public health in more than one area. For example, at one time megadoses of niacin were advocated for the treatment of schizophrenia, a disease for which there is no definitive cure, only symptomatic relief; but the results were too controversial for niacin to qualify as an antipsychotic drug—except for the pellagra-associated dementia. Finally, niacin is still used in very large doses as a medication for lowering blood lipids. Niacin lowers serum cholesterol, reduces LDL levels, and elevates HDL,* some of the factors generally associated with cardiovascular disease (*J. Am. Med. Assoc.* **231** [1975], 360). Multigram levels of niacin are needed to bring about these changes, so they are obviously drug effects of unknown mechanism, not related to a vitamin deficiency. Probably the appeal of niacin stems from the fact that it is a naturally occurring compound that does not need FDA approval, unlike the lipid-lowering molecules now manufactured in pharmaceutical laboratories.

For the sake of completeness, let me mention a rare familial disorder known as Hartnup's disease. Its symptoms are dermatitis, dementia, and sensitivity to light—that is, it is at least in part pellagra-like, and is also relieved by niacin or high-quality proteins. In distinction to pellagra, Hartnup's disease appears to be caused by defective metabolism of tryptophan, not by an inadequate dietary supply of this essential amino acid (Harvey L. Levy, *The Metabolic Basis of Inherited Diseases,* ed. Scriver et al., 6th ed., 1989, p. 2515).

Bibliography

1. K. J. Carpenter, ed. (1981), *Benchmark papers in biochemistry: Pellagra.* Stroudsburg, Pennsylvania, Hutchinson-Ross. (This valuable bibliography, now out of print, contains all major references to the pellagra literature up to 1981.)

2. F. Thiery (1932), Descriptions of a malady called mal de la rosa, in R. H. Major, *Classic descriptions of disease,* p. 575. Springfield, Illinois, Charles C. Thomas.

3. Sylvanus G. Morley (1947), *The ancient Maya,* rev. G. W. Brainerd, 2nd. ed. Stanford, Stanford University Press.

4. J. Goldberger and G. A. Wheeler (1915), Experimental pellagra in human subjects brought about by a restricted diet, *Pub. Health Rep.* **30,** 336.

5. J. Goldberger, G. A. Wheeler, and E. Sydenstricker (1918), A study of the diet of non-pellagrous and pellagrous households in textile mill communities in South Carolina in 1916, *J. Am. Med. Assoc.* **71,** 844.

*HDL is high-density lipoprotein; LDL, low-density lipoprotein.

6. J. Goldberger and W. F. Tanner (1922), Amino acid deficiency, probably the primary etiological factor in pellagra, *Pub. Health Rep.* **37,** 462.

7. C. A. Elvejhem, R. J. Madden, F. M. Strong, and D. W. Woolley (1937), Relation of nicotinic acid and nicotinic acid amide to canine black tongue, *J. Am. Chem. Soc.* **59,** 1767.

8. T. D. Spies et al. (1939), The use of nicotinic acid in the treatment of pellagra, *J. Am. Med. Assoc.* **110,** 662.

9. T. Jukes (1980), Conquest of pellagra, *Fed. Proc.* **40,** 1519.

10. K. J. Carpenter (1980), Effects of different methods of processing maize on its pellagragenic activity, *Fed. Proc.* **40,** 1531.

11. J. S. Wall, M. R. Young, and K. J. Carpenter (1985), Transformation of niacin-containing compounds in corn during grain development, *Fed. Proc.* **40,** 1282.

12. E. Kodicek et al. (1956), The effect of alkaline hydrolysis of maize on the availability of its nicotinic acid to the pig, *Brit. J. Nutr.* **10,** 51.

13. A. E. Harper, B. Punakar, and C. A. Elvejhem (1958), Effect of alkali treatment on the availability of niacin and amino acids in maize, *J. Nutr.* **66,** 153.

14. K. J. Carpenter, E. Kodicek, and P. W. Wilson (1960), The availability of bound nicotinic acid to the rat, *Brit. J. Nutr.* **14,** 23.

15. Y. Nishizuka and O. Hayaishi (1962), Enzymatic synthesis of niacin nucleotides from 3-hydroxyanthranilic acid in mammalian liver, *J. Biol. Chem.* **238,** 438.

11

Animal and Microbial Models

―――

The choice of animal models, sometimes exotic,

is important for biomedical research and

drug design.

Over the years, research in my laboratory employed a large variety of biological test systems, sometimes by rational choice and just as often by chance. Of course, early in my career before I became an independent investigator, it was the supervisor who made the decision. Later on the nature of the biochemical problems dictated the choices, ranging from the simplest microbes to the human organism. Biochemical diversity became one of my major interests.

Lipids of Tubercle Bacilli

I left the Technische Hochschule in Munich with a master's degree in chemical engineering. In actuality, my undergraduate training was largely in organic chemistry. I took only one chemical engineering course—on the "brewing process," appropriate to the locale.

Through the good offices of Hans Fischer, my professor, I found a temporary haven at the Schweizerisches Institut für Höhenforschung in Davos, a Swiss mountain resort. The head of the institute, Frederic

Roulet, was interested in the pathology of tuberculosis, a disease still rampant in the 1930s. One of my assignments* was to study the biochemical basis of an intriguing phenomenon. Lipid fractions obtained from human tubercle bacilli when injected subcutaneously into experimental animals caused epidermal tissue changes (tubercular granulomas and giant cells) indistinguishable from those elicited by injection of live or heat-killed bacteria. Roulet's interests were twofold. First of all, the biologically active principles in the lipid extracts were to be characterized chemically. Second, he felt that the species specificity of these lipids (human versus bovine) was worth investigating. At the time, bovine tuberculosis was quite widespread in some Swiss cantons.

I set about purifying the active principles from cultures of human and bovine bacilli. Both strains were grown on a large scale in Zurich at the university's Department of Bacteriology. I collected the cells, washed them with organic solvents, and carried the bacteria back to Davos in a large Dewar flask (a thermos bottle invented by the British physicist Sir James Dewar). I did not reveal the contents of the flask to my fellow train passengers.

Tests showed that the solvent-extracted tubercle bacilli were still viable! Extracts from both strains yielded active fractions which I identified as phosphatides, indistinguishable by chemical, admittedly rather primitive, criteria. Only a biological test could answer the question of species specificity. What caused me to volunteer as the experimental animal I do not recall—perhaps it was expected of the young assistant. (The practice was not unusual in early biomedical research.) At any rate, the lipid material from the human strain was injected intradermally into my left arm, and the bovine phosphatide into my right arm. Only the human fraction gave a positive response, to Dr. Roulet's delight. As far as I know, the presumably subtle chemical differences determining the species specificity of the human and the bovine phospholipids have never been established. The scars on my arms resulting from surgical removal of the pimples are still highly visible—after fifty-five years, a reminder of youthful foolishness.

*My first research at the institute was to resolve the controversy over whether or not tubercle bacilli contained cholesterol. This project is described in Chapter 2.

As a graduate student admitted to the biochemistry department of Columbia Medical School in 1936, my research leading to the Ph.D. degree was straightforward organic chemistry. Biochemistry was still to become my chosen field. It did when Rudolf Schoenheimer, the pioneer in introducing stable isotopes for metabolic studies, took me on as a postdoctoral student. Later on, he confessed that he hesitated to hire me because my dissertation was so "thin," an eleven-page paper in the *Journal of Biological Chemistry* at a time (1938) when the journal's format was small, the print large, and the margins wide.

For tracing metabolic conversions with the aid of isotopic tracers, rats or mice were customarily used at the time. The choice of small animals was dictated by the scarcity of stable isotopes, especially ^{13}C, ^{15}N, and ^{18}O. Schoenheimer's laboratory was privileged to receive these isotopes, not yet available commercially, from Columbia's chemistry department through the generosity of Harold C. Urey, the discoverer of these stable isotopes.

Only in two instances were larger animals realistic choices. For demonstrating the conversion of cholesterol to bile acids a dog was subjected to a surgical procedure known as cholecystonephrostomy, a technique diverting bile into the urinary tract. The isolated bile acids were shown to be labeled after intravenous administration of labeled cholesterol, a suspected but not yet proven metabolic transformation. The scarcity of labeled cholesterol did not permit a duplicate experiment.

Restrictions other than size of the experimental subject became a serious problem when I planned to demonstrate the precursor role of cholesterol for steroid hormones. In the animal body, hormones ordinarily occur in minute amounts, not sufficient for isolation and chemical analysis. To take advantage of the relatively massive urinary excretion of pregnanediol, a progesterone metabolite, in the late stages of human pregnancy was a more promising if unusual approach. To quote from my 1945 paper: "Deuterocholesterol was ingested by a woman in the eighth month of pregnancy, the progesterone metabolite isolated from the urine excreted during the experimental period in the form of pregnanediol glucuronidate, and analyzed for deuterium. Significant isotope concentrations were present in the samples showing clearly that pregnanediol had been formed directly from cholesterol." One month later my son Peter was born, an event that identified the essential collaborator in the experiment.*

From "Intact Animal" to "In Vitro" Experiments

When a test compound is fed to an animal or injected, the site of conversion to suspected product cannot be specified, let alone quantified. In essence, the intact animal is a "black box." Perfusion of isolated organs such as liver or kidney was a step in the right direction, but the procedure was technically complex and not always reproducible. As early as 1923, Otto Warburg introduced the tissue slice procedure. Hans Krebs improved it. Many classic experiments on respiration, the urea cycle, and the citric acid cycle owed their success to this first in vitro methodology. Tissue slices contain largely intact cells, which are impermeable to most coenzymes including ATP, and therefore do not support reactions that are energetically uphill. This barrier was overcome in 1946 by Nancy Bucher's design of "loose pestle" homogenates (biologically active cell-free preparations), a boon to those interested in biosynthetic mechanisms (see also Chapter 7). In essence, the Bucher loose pestle homogenizer enabled the experimenter to fractionate intracellular elements and to separate enzymatically active from inactive materials. This was of course not possible with intact cells. Our own research on the formation of fatty acids, cholesterol, and peptide bond formation, reactions requiring ATP as an energy source, greatly benefited from the technique introduced by Bucher.

Domestication of the Rat

Rats used for experimental purposes today are descendants of the Norway rat that originated in Siberia and spread over Europe and other

---

*Before involving my family in the experiment, I had attempted to persuade the Department of Obstetrics and Gynecology to provide me with a suitable patient. The request was curtly denied.

continents with the development of commerce (Castle, 1947). At some unspecified time albino mutants probably appeared on the scene and eventually became the earliest domesticated breed. The first recorded breeding experiments were undertaken by H. Crampe in 1880. White rats of European origin were bred at the University of Chicago by H. Dean, who transferred his stock to the Wistar Institute, a source of laboratory colonies for several decades, later replaced by the commercially available Sprague-Dawley breed. For identification of vitamins, essential amino acids, essential fatty acids, and minerals, these inbred strains became the standard animal model. With a few minor exceptions, these requirements could be extrapolated to the nutritional needs of humans. The same is true to a large extent of the multitude of metabolic pathways mammals employ. Today the rat model remains popular as the cheapest mammal by far. Moreover, opponents of all animal experimentation seem to be little concerned about rodents. Their concern is, or perhaps used to be, to protect their household pets. But as a premier animal model, the rat has lost much of its significance. (This is a personal opinion, not shared by all investigators.)

For example, the extensive use of rats as models of carcinogenesis in humans, though still widespread, has become increasingly problematic. This is true for all animal species having a much shorter life span than humans. To demonstrate carcinogenic effects in small animals, abnormally large doses have to be given for brief periods in order to "compensate" for the age factor. It is doubtful that such extrapolations are valid, because pharmacokinetics, the rate of carcinogen removal or transformation, is dose related. Moreover, the biochemical mechanisms for carcinogen activation or detoxification may differ unpredictably among species.

Not All Rats Are Alike

At this point I must mention an episode during my stay at the University of Chicago that caused me considerable anxiety, at least temporarily. Stimulated by some earlier observations implicating a role of the hormone insulin in the conversion of carbohydrate to fat, we tested the incorporation of labeled acetate into long-chain fatty acids by rat liver in vitro. We were delighted to find that insulin enhanced this process about fourfold. It was the first demonstration of an insulin effect in isolated liver. The results were published promptly in the *Journal of Biological Chemistry* (Bloch and Kramer, 1948).

Subsequent attempts to repeat and confirm the original findings failed utterly. Various batches of insulin obtained from drug companies, Eli Lilly as well as Squibb, were tested to no avail. I drafted a retraction letter to the journal, but before mailing it I went to the animal room in the university's Abbott Laboratories. When I asked the caretaker whether something might be wrong with our Chicago rat colony, he said, "Oh, we no longer breed our own. We now buy Sprague-Dawley rats from the Charles River Laboratories. They are so much easier to handle; they never bite." I immediately procured rats of various strains, among them Wistar and Yale, still in use at the time in a number of laboratories. Only with liver slices from the Yale rats were we able to confirm our already published results. An older departmental colleague seemed to remember the history of the Chicago-bred rats: they were the offspring of the Yale strain, slightly diabetic (insulin deficient), and of unpleasant temperament. I did not have to mail my letter of retraction.

Today it is, or should be, routine to identify the genetic ancestry of inbred experimental animals—easy enough with rats and mice but impractical (or impossible) with less conventional mammalian species.

My largest experimental animal ever was the shark. By 1952 our research on cholesterol biosynthesis pointed very strongly to the intermediacy of squalene, a thirty-carbon branched-chain hydrocarbon so named because it was first isolated in large amounts from the livers of members of the shark family (*Squalidae*).* How was one to obtain radio labeled squalene in order to show its metabolic relationship to cholesterol? I

---

*The shark's liver accounts for 30 percent of the animal's body weight, more than three times that of other animal livers. Moreover, the organ has the consistency of blubber owing to its abnormally high content of squalene and related hydrocarbons. Native Bermuda fishermen derive lamp oil from shark livers simply by squeezing the oily tissues.

considered working with dogfish, the smallest member of the shark family, a project that seemed feasible in the well-equipped Marine Biological Laboratory at Woods Hole. Still, the thought occurred to me that an even more attractive locale for summer research might be the Bermuda Biological Station. My family agreed. In response to my inquiry, the director of the Bermuda laboratory gave assurances that I would be welcome to carry out the research I proposed.

On arrival in Bermuda, I learned that the only dogfish ever captured around the islands was housed in the local aquarium. For obvious reasons, this specimen was unavailable for experimental purposes. Could we, if lucky, catch a baby shark of manageable size? On numerous fishing trips we encountered only the fully grown (6–8 feet) variety, but in one instance a baby was hooked* (weight about 100 pounds) and successfully hauled into the largest fish tank available at the station. The experimental plan was to excise the liver, prepare tissue slices, incubate them with ^{14}C acetate, and isolate squalene. Alas, the shark liver yielded only a greasy mess instead of slices. The scrapings from the slicing board contained no radioactivity.

Meanwhile, Robert Langdon, a physician working on his doctoral dissertation in the Chicago laboratory, used the more realistic but less spectacular approach of feeding radioactive acetate to rats, along with unlabeled squalene. His experiments afforded labeled squalene and hence proved the biosynthetic sequence acetate $\rightarrow$ squalene $\rightarrow$ cholesterol, needless to say at considerably less expense than my procedures.

Once we had shown that acetic acid is a carbon precursor of cholesterol (1942), the question arose whether this simple two-carbon compound was a major or a minor source. Cholesterol contains twenty-seven carbon atoms, and if acetate provided them all at least fourteen of these molecules had to engage in the building process. For various reasons, among them

*Shark fishing was not a pleasant experience. A local fisherman and I went to the island's shore where garbage was dumped. A big hunk of tuna was the bait hanging from a large meat hook. Within a few minutes there was an unmistakable bite.

the black box paradigm, this issue could not be resolved at the time by whole-animal experimentation.

Genes Act by Regulating Definite Chemical Events

An experimental tool for answering questions such as the one just cited was introduced in 1941 by Beadle and Tatum, discoverers of the biochemical mutant technique. By exposing the mold *Neurospora crassa* to x-rays, they were able to delete or inactivate single genes responsible for the synthesis of selected molecules. One of the *Neurospora* mutants was blocked in the metabolic degradation of glucose or fat to acetic acid and consequently required external acetic acid for growth. In collaboration with Ed Tatum and his group, Irving Zabin in our laboratory grew the mutant strain on a medium containing isotopically labeled acetic acid. Since the mutant cells themselves were unable to produce acetate, products of acetate metabolism had to be derived entirely from the external isotopic acetate. The cellular sterol isolated from the *Neurospora* mutant, when analyzed, contained nearly the same abundance of isotope as the acetate added to the growth medium, thereby proving that acetic acid was the sole carbon source for the sterol structure.

This knowledge greatly simplified our task in one important respect. A search for *ultimate* precursors of cholesterol was no longer needed. Still, the major challenge remained: how the numerous acetate units assemble to form the twenty-seven-carbon sterol molecule. As it turned out, the overall process involved some thirty discrete steps and intermediates.

A chance observation, made in the pursuit of totally unrelated objectives, critically advanced the problem of key intermediates in sterol biosynthesis. In the mid-1950s the Merck, Sharp & Dohme laboratories were engaged in a major search for novel vitamins. I suppose their motivation and aim was to find bacterial growth factors or vitamins that might be useful, if not necessarily essential, for human nutrition. A number of *Lactobacillus* mutants (*Lactobacillus acidophilus*) were screened, and among them a few were found to be "acetateless." Like the *Neurospora* mutant, they had an absolute requirement for acetate in order to grow. In their search for "acetate-replacing factors," the Merck group found rich sources of the desired activity in various biological materials (for example, "distillers' dried soluble," trade name Hiram Walker's "stimuflow"), ultimately identified as the six-carbon compound mevalonic acid, or MVA. This molecule, the acetate-replacing factor, was an exceedingly effective precursor of cholesterol, later shown to be derived from three molecules of acetate (Tavormina et al., 1956).

The discovery of mevalonic acid as a key intermediate in squalene and cholesterol biosynthesis from acetate is a classic example of serendipity, the more remarkable because at the time the function of MVA in the bacteria that require it was unknown. Sterols do not occur in these organisms. Only a decade later was it recognized that MVA is a key intermediate for a great variety of "polyisoprenoids," which include, apart from cholesterol, certain vitamins, coenzymes, natural rubber, and various terpenes.*

Shortly after the Merck group had reported the discovery of MVA, I met the senior research director, Karl Folkers, and asked for his consent to explore the further metabolism of MVA to cholesterol, expecting a negative reply. Merck had every right to exploit the discovery. To my great surprise, Folkers not only failed to object but generously offered me samples of MVA for whatever experiments my group was planning.**

All along we had hoped to identify additional natural mutant organisms

---

*Terpenes, found primarily in plants and in some bacteria, were thought to be derived from MVA by way of the five-carbon intermediate isopentenylpyrophosphate (IPP). Surprisingly, an alternative route to IPP was discovered recently that did not involve MVA (Rohmer et al., *Biochem. J.* **295** [1993], 517; D. Arigoni et al., private communication, 1994). This novel pathway, not yet known in detail except for the fact that it starts with pyruvate rather than acetate, is found in the terpenes of some bacteria and in the terpenes of the gingko tree, a further example of an alternative mechanism for generating independent pathways leading to a common intermediate.

**The discovery of MVA ultimately proved to be a boon for the pharmaceutical industry. Metabolic control of cholesterol synthesis in mammals occurs at the level of MVA formation. A search of inhibitors of this process has yielded MVA analogs, drugs effective in lowering blood cholesterol.

that were blocked at some stage of cholesterol biosynthesis and might accumulate intermediates that other animal species convert to sterols. When we learned that insects had vitamin requirements not seen in vertebrates, our search appeared successful. R. P. Hobson at the London School of Hygiene and Tropical Medicine, searching for insect vitamins in 1936, was the first to obtain evidence that one insect growth factor might be a sterol. The insect physiologist Gottfried Fraenkel confirmed this finding for the hide beetle *Dermestes vulpinus,* an obligate carnivore. Fraenkel had moved from Britain to the University of Illinois at a time when our group in Chicago was anxious to investigate the basis of this nutritional requirement with the aid of isotopic tracers. I wrote to Fraenkel proposing a collaboration.

Our work together went smoothly and produced a modest paper that confirmed the absence of sterol synthesis and of identifiable sterol precursors in *Dermestes vulpinus.* Still, we failed in our objective to identify the nature of the genetic deficiency, why sterols are essential nutrients for insects. The sterol-derived molting hormone ecdysone had not yet been discovered. There was some suspicion that the block occurred at the squalene stage, but this could not be verified. Apparently the missing genes or enzymes remain unknown to this day. Still, our later work with insects paid other dividends, notably in teaching us how carnivorous and herbivorous insects handle plant sterols (see Chapter 2).

Going Back to School

Except for the occasional excursions mentioned, my early research had been limited to rats, whole animals, and isolated liver preparations. During the 1950s biochemists, many among them middleaged, began to seize the opportunities provided by the remarkable developments in molecular biology. The use of microbial mutants, introduced by Beadle and Tatum with the mold *Neurospora,* had been extended to bacteria with the aid of the elegant penicillin technique. Joshua Lederberg and Bernard Davis discovered it independently in 1952. *Escherichia coli* became the organism of choice for biochemists whose primary interests were biosynthetic processes. I was particularly impressed by the elucidation of the pathways to the aromatic amino acids tryptophan, tyrosine, and phenylalanine (Davis, 1956), totally unknown previously and an outstanding example of the power of the mutant technique.

In order to ease the transition from familiar to unfamiliar biological systems, I decided in 1957 to go back to school. I had been told of a popular microbiology summer course taught by C. B. van Niel at the Hopkins Marine Station in Pacific Grove, California—located on the Monterey Peninsula, an added attraction. My application to enroll in the course was accepted, along with those of some fifteen biochemists, all anxious to be introduced to microbiological techniques (the average age of those in the class was forty-two). We heard a rumor that applicants who had already taken a microbiology course elsewhere were not admitted.

The exceedingly demanding course taught me important lessons that were to influence and redirect my research from then on. Van Niel, a classic bacterial physiologist in the tradition of the Dutch school of Beyerinck and Kluyver, first made the class aware of the rich variety of microorganisms ("beasties," he called them) and their diverse lifestyles. Second, we learned from van Niel that Nature provides the investigator with numerous organisms to choose from, some uniquely suited for studying a specific biological phenomenon.*

Van Niel was best known for his unifying concepts of photosynthesis, a subject he discussed in great detail. I am fairly certain that his lectures plus the various contacts with my classmates acquainted me with the phytoflagellate *Euglena gracilis* and its dual lifestyle. In the light, these organisms are green protists, deriving their energy by photosynthesis as do higher plants. When grown in complete darkness, they lose their photosynthetic apparatus. They turn colorless (etiolated) and adopt animal-like metabolic patterns. These lifestyles are readily reversible. *Euglena* was therefore an organism of choice for studying the process of chloroplast

---

*Classic physiologists and biochemists working almost exclusively with higher animals had already made some of the obvious choices: pigeon breast muscle, which has the highest rate of oxygen uptake and was the richest source of respiratory enzymes (A. Szent-Györgyi); or mammalian liver, the exclusive site for converting protein nitrogen to urea. H. A. Krebs discovered the urea cycle in liver.

development. Our interests were not in photosynthesis per se, but in the chemistry of membranes, especially their fatty acid composition. We could show that the lipid patterns expressed in the dark and in the light were those typical of animals and plants respectively and could be changed at will by growing *Euglena* either in the light or in the dark. (I have mentioned *Euglena* in an earlier chapter on Alternative Pathways.)

Another piece of helpful information van Niel mentioned—almost casually—in his lectures benefited me perhaps more than any other for planning future research. The class was told that common brewer's yeast, *Saccharomyces cerevisiae*—the organism that led to Pasteur's fundamental discovery,* which he termed "la vie sans air"—is in fact microaerophilic, not strictly anaerobic. This information came from a 1954 paper by Andeassen and Stier in the *Journal of Cellular and Comparative Physiology,* a periodical then unfamiliar to me. These authors noted that in the *strict* absence of oxygen, yeast fails to grow unless supplied with cholesterol and an unsaturated fatty acid. Obviously oxygen, albeit at very low atmospheric pressures, is essential for the biosynthesis of these lipid molecules. This realization was essential to our later research and also stimulated my interest in the role of oxygen in the evolution of biochemical pathways and organisms.

Mycoplasmas, Bacteria without Cell Walls

On a return trip from a Gordon Conference in the 1970s, my fellow passenger Shlomo Rottem, a visitor from the University of Jerusalem, acquainted me with mycoplasmas, bacteria he regarded as ideal for investigating the role of cholesterol in membrane function. He was an expert in the field. I was vaguely aware of these organisms, but since some

---

*I must mention here Pasteur's motivation for studying brewer's yeast fermentation. In 1871, following the defeat of France in the Franco-Prussian War, he stated, "I will do my utmost to restore the glory of France" (André Lwoff, private communication). Realizing that while French wines were the best in the world, French beer did not stand comparison with the products of the German brewing industry, Pasteur set out to correct that deficiency.

mycoplasmas are pathogens, I was reluctant to introduce them into the laboratory. Rottem told me of one harmless species, *Mycoplasma capricolum,* and how easy it is to grow, provided cholesterol is included in the growth medium.

In fact, mycoplasmas are the only exception to the rule that bacteria, in contrast to eukaryotes, neither require nor synthesize sterols of any kind. One can rationalize this exceptional nutrient requirement. What distinguishes mycoplasmas, also known as mollycutes (soft shelled), from all other bacteria is the absence of the otherwise universal peptidoglycan cell wall, a rigid envelope that enables bacteria to retain their shape and structural integrity in hostile environments. Mycoplasmas, lacking this protective coat, use cholesterol instead as a membrane stabilizer. When free living, that is, as infectious agents, they derive it from a number of cholesterol-rich animal tissues; they are parasites. For that reason mycoplasmas are feared by biologists who work with animal cells in culture.

A recent news report mentioned that mycoplasmas might contaminate cultures of the AIDS virus (HIV), one possible explanation for the controversy surrounding the identity of the original isolates of the virus. Such mycoplasma contaminations are difficult to control with antibiotics, especially of the penicillin type. These antibiotics interfere specifically with bacterial cell wall synthesis; they are not active against the wall-less mycoplasmas.

The above examples illustrate why biochemists settle on one test system in preference to others. The choice may be a matter of convenience or an accidental encounter. In my own case, the additional motivation was to find out how different cells or organisms solve a given functional problem at the biochemical level (Table 11.1).

Early Animals Models

So far this chapter has chronicled the diverse biological systems chosen because they proved helpful, if not necessarily essential, in investigating the mechanism of sterol biosynthesis. What follows are comments on two model organisms, inspired probably by accidental reading without any

Table 11.1
Biological systems used in the Bloch laboratory

Organisms	Date
Human and bovine tubercle bacilli	1934–1935
Rats and mice; intact-animal experiments	1939–1946 and after
Dog	1943
Pregnant woman	1945
Rats, in vitro; tissue slices and extracts	1946
Pigeon liver	1949
Yeast (*Saccharomyces cerevisiae*)	1950–1989
Neurospora crassa	1951
Shark liver	1952
Insect larvae:	
Hide beetle (*Dermestes vulpinus*)	1959
Florida cockroach (*Blatella germanica*)	1959
Tetrahymena corlissi, a protozoan	1961
Clostridium butylicum, anaerobic bacteria	1961
Mycobacterium phlei (smegmatis)	1962
Ciliated protozoa	1963
E. coli	1963
Blue-green algae	1964
Euglena gracilis, Anabena variabilis (phyloflagellates)	1964
Micrococcus lysodeicticus	1963
Torulopsis utilis (aerobic yeast)	1963
Mycoplasma capricolum	1978
Yeast mutant GL-7	1980

explicit motive to further my own research. It goes without saying that here I will be trespassing on less familiar territory.

The first researches on experimental atherosclerosis illustrate why the choice of an animal model is crucial to success or failure. Early in the twentieth century two Russian pathologists first demonstrated that degenerative diseases resembling human atherosclerosis are inducible by dietary means. Rabbits were their choice as experimental animals, a common selection at the time, presumably because they propagate rapidly and have a gentle disposition. A. Ignatowski (1909), feeding his rabbits a diet rich in meat, milk, and eggs, produced such arterial lesions, while N. Anitschkow (1913) saw the same fatty deposits, or plaques, in rabbit blood vessels after administering large amounts of cholesterol. Thus the first (causative?) link between dietary cholesterol and cardiovascular disease was experimentally established. Most physicians still hold this view, though other factors, genetic and behavioral, undoubtedly play a major role as well in human atherosclerosis.

The etiology of this degenerate disease is "multifactorial." Why rabbits were chosen as experimental animals by the Russian pathologists is an intriguing question. In retrospect, two comments are relevant. First, rabbits are obligate herbivores; thus cholesterol is not one of their normal dietary constituents. They may have a limited capacity for metabolizing cholesterol and therefore tend to deposit excessive amounts in their arteries. Why did Ignatowski and Anitschkow not use rats or other rodents instead? The answer is very simple. Before 1912, when inbred rat strains were first raised, only wild rats were available for experimentation.

In a sense, the early choice of rabbits instead of rats was fortunate. In later experiments (1939) Anitschkow himself realized that the response of the rat strains then available differed markedly from that of rabbits. In rats or other rodent species, dietary manipulation induces neither hyperlipidemia nor atheromatous lesions. Not until the 1950s were procedures developed for lowering the rat's resistance to atherogenesis by chemical or genetic devices (Wissler et al., 1952). Drugs such as thiouracil, that depress thyroid activity, interfering with the iodination of thyroxine precursors in the thyroid gland, were especially effective in rendering rats sensitive to lipid overloads. Also, two inbred rat strains, the spontaneously hypertensive SHR and the BB prone to developing diabetes mellitus, became available and potentially useful for defining the etiology of cardiovascular disease.

Along these lines, dramatic developments have recently taken place (Lawn et al., 1992). The novel technique for transferring genes from one species to another has been shown to render mice, and presumably rats as well, susceptible to atherosclerosis and its various manifestations. In humans, high plasma levels of a protein known as apolipoprotein, Lp(a), constitute a principal factor associated with the disease. The human gene that expresses this blood protein has been isolated, cloned, and transferred to otherwise normal mice. These transgenic mice do develop atherosclerosis when given high-fat diets. As these startling findings promise—and there will undoubtedly be others—the search for valid, relevant animal models for studying human disease may soon come to an end.

Nevertheless, the normal rat, especially the widely used Sprague-Dawley strain, remains a popular animal model for studying the relation of human diet and atherosclerosis. For the reasons just given, the relevance of such information is questionable. During the last decade the major advances in the understanding and etiology of atherosclerosis have been made with human cells in culture rather than with whole animals. This technique, more than any other, has pinpointed the principal defects in human familial hypercholesteremia (Goldstein and Brown, 1977; Brown and Goldstein, 1984).

An Exotic Model for Biomedical Research

"The nine-banded armadillo (*Dasypus novemcinctus*), a primitive animal, has unique potential for use in many areas of medical and biological research. Among characterizations which make this animal a useful research model are low body temperature (32°–35° C) and a regular production of litters of quadruplet young and a long lifespan (ten to fifteen years). It was because of the low body temperature that I first became interested in the armadillo as an animal model in leprosy studies, since it was known that *Mycobacterium leprae* multiplies best in man in the cooler regions of the body."

Thus wrote Eleanor Storrs (1982), a biochemist at the Gulf South Research Institute in New Iberia, Louisiana. At the same time, she and W. F. Kirchheimer successfully transferred the infectious agent causing human leprosy into this exotic mammal. A few years later, investigators at the same research institute discovered indigenous leprosy in an armadillo captured in the wild. Indigenous leprosy has since been observed in two other animal species, a chimpanzee and a mangabey monkey, but the disease is rare in these primates. Storrs points out that the armadillo has the additional, apparently unique property of producing quadruplets, four replicate (identical) offspring.

Leprosy, Hansen's disease, is described as a chronic infection caused by *Mycobacterium leprae,* an organism with high infectivity but low pathogenicity. Following Hansen's discovery of the leprosy bacillus in 1876, the disease was found to be distributed worldwide but concentrated in equatorial regions around the globe, in Africa, Southeast Asia, and South America, with a total of 12 million to 20 million recorded cases. In North America only a few thousand affected individuals remain, nearly all in leper colonies in Louisiana and sporadically in southern Texas. With the advent of recent antibacterial agents (sulfones, rifampines) the number of lepers and leper colonies has steadily declined. It may come as a surprise that *M. leprae* was the first infectious bacterial agent to be isolated and one of the last to yield to antibiotic therapy. While leprosy appears to be no longer a major public health problem, the recent resurgence of tuberculosis—also caused by a mycobacterium—should be a warning. Acquired drug resistance remains a constant threat to the management of infectious diseases.

Storrs's awareness of the armadillo's low body temperature was a key element in her decision to use this species as an animal model for human leprosy. Usually, warm-blooded mammals are homeotherms. They maintain their internal body temperature at 37° C. Knowing that the optimal growth temperature for *M. leprae* cultures lies in the same low range (32–35° C) as in the infected animal's host, she reasoned that the armadillo's disposition to the disease could be explained because it provided an ideal, perhaps unique, environment for the invading bacterium. Clinical diagnosis had located the diseased regions in the cooler, exposed parts of the human body, ulcerations occurring especially in the skin, leading to disfigurement of fingers ("mitten hand"), lesions in face and ears, lower lip, eyebrows, and peripheral (cutaneous) nerves. The bacterial predilection for these cooler, exposed regions of the human body precisely matches the location of the inflicted regions in the armadillo. In fact, the human leper's skin resembles in appearance that of the dorsal leathery carapace, the chitinous shield covering the back of the armadillo.

How is leprosy transmitted from a disease carrier to a healthy organism, and especially from one species to another? It has long been believed that leprosy in humans is transmitted solely by human carriers. Recent evidence points to animal reservoirs as well, especially armadillos, and less frequently certain monkeys that must be considered vectors. In Louisiana, among one thousand armadillos captured in the wild, 10 percent have been found to be afflicted with indigenous leprosy. The habitats of these animals in Louisiana are swampy, dense forests harboring large populations of insects.

Histological, bacteriological, and immunological tests all show that the captured armadillos are identical in these respects to those inoculated in the laboratory with *Mycobacterium leprae*. Leprosy therefore appears to be transmissible within the same species. Since armadillos have not been bred successfully and domesticated armadillos do not exist, it seems unlikely that they acquired the disease directly from leprosy patients. Transfer from one animal to another could be by inhalation or via the mother's milk. Insects are the major source of the armadillos' food in their natural habitat, so insect vectors are another likely, but not proven, transmission path.

As for the ultimate origin of human leprosy, the consensus appears to be that armadillos in the wild are the likely source of infection. Geographically, the occurrence of the disease in man and beast overlaps.* More conclusively, the size of the genome is the same and DNA homology is 100 percent for *M. leprae* isolated from both species.

Storrs mentions earlier studies (1963–1965) on the effects of the notorious teratogenic drug thalidomide on armadillo embryos. The drug is nontoxic for most adult experimental animals but produces congenital transformations in armadillo offspring when administered to pregnant females. Storrs therefore proposed the use of this exotic mammal as a model for studying potentially teratogenic drugs. There is no evidence that her advice was followed.

The history of thalidomide and its ultimate withdrawal from the market highlights the problems faced by pharmaceutical companies prior to preclinical drug testing. Often the choice of animal models is fortuitous rather than rational. Thalidomide was synthesized in 1957, and the patent assigned to Chemie Gruenenthal in Berlin.** The drug was developed as a sedative with hypnotic properties and quite widely used in Europe. In the

---

*An apparent absence of armadillos in Hawaii would contradict this correlation. Perhaps immigrants, infected elsewhere, brought leprosy to the islands.

**I worked in the firm's laboratory during the summer of 1935. It was a reputable pharmaceutical company, owned by distant relatives of mine.

United States, thalidomide never reached the public; the Food and Drug Administration would not approve its use. A skeptical FDA commissioner, Frances Kelsey, instituted drug testing of pregnant animals—not a previous practice—and discovered the disastrous thalidomide effects on embryonic limb development. They are manifest not only in human embryos but also in pregnant rabbits and monkeys, but not in rats.* The unfortunate fact that thalidomide slipped through the net of regulatory agencies in many Western countries led to the tragic birth of thousands of malformed infants.

Thalidomide was one of the most dramatic setbacks ever for a drug company. My personal belief is that insufficient emphasis is given inter alia to the diversity of drug metabolic pathways and hence to potential toxicity in various animal species. For example, in responsive systems such as developing human, rabbit, and monkey embryos, but not in the human adult and rats, thalidomide is converted oxidatively to so-called arene oxides. These derivatives can bind irreversibly to DNA. The result is gene mutation leading to teratogenesis, or "monstrous" malformation.**

Few would argue against the concept that the ideal model for the human organism is man. Perfected techniques for growing human cells in culture have brought this goal a step closer. Researches on human familial hypercholesteremia are an outstanding example. Yet as a general methodology, even cultures of human cells or organs present problems. Which of the numerous cell types is one to choose? The genetic background of the human race is diverse—how diverse will probably remain unknown until the human genome project is completed.

For the time being we have to accept the fact that no medication is or will be 100 percent effective, with the exception of vaccines. For the

---

*Once again, the "normal" rat disqualifies as an universal animal model, as we saw earlier in the case of atherosclerosis.

** Decades later, thalidomide's reputation has been dramatically changed. Research in the laboratory of G. Kaplan (Rockefeller University) has demonstrated that the infamous drug suppresses the activation of latent human immunodeficiency virus type I (HIV-I) in certain cell lines and blood cells of patients with advanced HIV-I infection and AIDS. (S. Makonakawekeyoon et al. (1993) Proc. Nat. Acad. Sci., USA **90**, 5974, p. 23).

pharmaceutical industry, preclinical testing of drugs in several animal models for the time being remains the only alternative. It is for this reason that the virtues of comparative biochemistry have been stressed.

Bibliography

1. K. Bloch (1945), The biological conversion of cholesterol to pregnane-diol, *J. Biol. Chem.* **157**, 611.

2. W. E. Castle (1947), The domestication of the rat, *Proc. Nat. Acad. Sci.* (USA) **33**, 109.

3. K. Bloch and W. Kramer (1948), The effect of pyruvate and insulin on fatty acid synthesis *in vitro, J. Biol. Chem.* **173**, 811.

4. G. W. Beadle and E. L. Tatum (1954) Genes act by regulating definite chemical events, *Les Prix Nobel,* Nobel Foundation. Stockholm, Imprimerie Royale, P. A. Norstedt.

5. L. D. Wright et al. (1956), Isolation of a new acetate-replacing factor, *J. Am. Chem. Soc.* **78**, 5273.

6. P. A. Tavormina et al. (1956), The utilization of mevalonic acid in cholesterol biosynthesis, *J. Am. Chem. Soc.* **77**, 4498.

7. B. D. Davis (1954–55). Biochemical exploration with bacterial mutants. *The Harvey lectures,* **50**, 230. New York, Academic Press.

8. A. Ignatowski (1909), Ueber die Wirkung des tierischen Eiweisses auf die Aorta, *Pathol. Anal. Physiol.* **198**, 248.

9. N. D. Anitschkow (1913), Ueber die Veränderungen der Kaninchenaorta bei experimentaller Arteriosklerosen, *Beitr. Pathol. Anat.* **56**, 379.

10. R. W. Wissler et al. (1952), The production of atheromatous lesions in the albino rat, *Proc. Inst. Med., U. Chicago,* **19**, 79.

11. P. A. Lawn et al. (1992), *Nature* **360**, 670.

12. J. L. Goldstein and M. S. Brown (1977), The low density lipoprotein pathway and its relation to atherosclerosis, *Ann. Rev. Biochem.* **46**, 897–930.

13. M. S. Brown and J. L. Goldstein (1984), A receptor-mediated pathway for cholesterol homeostasis, *Sci. Am.* **251**, 58–66.

14. E. Storrs et al. (1977), Leprosy in the armadillo: A model for leprosy research, *Science* **183**, 851.

15. E. Storrs (1982), The astonishing armadillo, *National Geographic* **161**, 820.

12

Carnivores, Herbivores, and Omnivores

———

"Dinah is our cat. And she is such a capital one for

catching mice, you can't think. And oh, I wish you could

see her after the birds!" The metabolic basis of obligate

carnivorous lifestyles is illustrated by four genetic

(enzymatic) deficiencies afflicting members of the

cat family. The mystery of the giant panda's

nutrition is probed.

The nutritional lifestyles animals practice are of three kinds: carnivorous (as keenly observed by Alice in Wonderland), herbivorous, and omnivorous. The human organism is the most versatile, capable of subsisting and reproducing by any of the three modes. Among animal species obligate carnivores, known behaviorly as predators, are at one extreme, while obligate herbivores, such as ruminants, are at the other. These lifestyles obviously have a biochemical or nutritional basis. We now know all of the essential molecules that the animal body fails to synthesize but must derive from food: the vitamins, essential amino acids, and some unsaturated fatty acids. Omnivores, man, and omnivorous animal species have a wide choice. They can meet these requirements either from plants or from the meat of other species. By definition, obligate carnivores and obligate herbivores do not have this choice. Their diets are a matter of necessity, forced on them by their genetic inheritance.

This chapter will deal primarily with obligate carnivores, animals that rely on the flesh of other animal species they kill or scavenge.

Carnivora

Carnivora constitute about 10 percent of all mammalian genera. Webster defines carnivores as "a mammalian order, *most* of which are *largely* or entirely carnivorous in habits. Their teeth are modified for a carnivorous diet, large and long to hold prey, sharp-edged incisors, molars; strong and thick for breaking bones, simple stomach (compared to herbivores). The two suborders are fissipeds (separated toes) and pinnipeds (fully webbed digits)."

I italicize *most* and *largely,* to emphasize that not all carnivores are *obligate* flesh eaters. Indeed, the zoological literature (Gittleman, 1989) lists the giant panda and the black bear as "strictly herbivorous carnivores." This convention of classifying an order according to dentition—the number, kind, and arrangement of teeth—among other anatomical features, but excluding nutrition, must strike the outsider as a paradox or an oxymoron. To the biochemist, and certainly in the popular mind, carnivores including mammals, and predatory fishes and birds, are necessarily meat eaters. "Raptor," a less common and more restrictive term, refers specifically to birds of prey (hawks, eagles, and the like), all of which appear to be obligate carnivores.

Traditionally, the vast scientific literature on the order Carnivora deals with behavioral characteristics, ecology, and morphology. Whether by choice or more likely of necessity, behavioral attributes have for the most part been studied with animals in the wild. As for the genetic basis of the carnivorous habit, precious little is known, the exception being the domesticated cat. Nutritional data on domesticated dogs are available, but they do not mention genetic defects, if any, of Canidae in the wild.

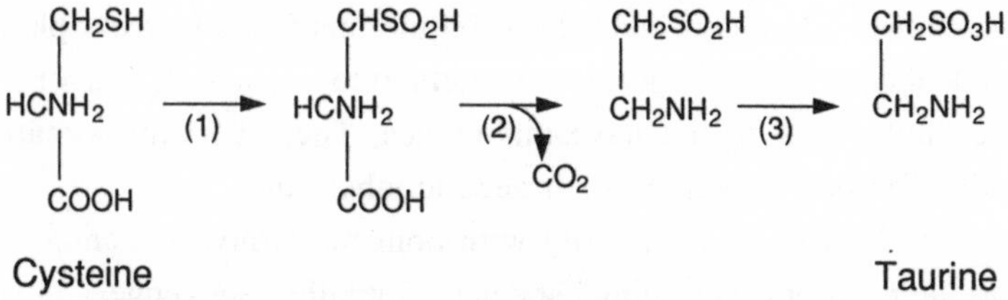

CH$_2$SH   CHSO$_2$H   CH$_2$SO$_2$H   CH$_2$SO$_3$H

HCNH$_2$ $\xrightarrow{(1)}$ HCNH$_2$ $\xrightarrow[CO_2]{(2)}$ CH$_2$NH$_2$ $\xrightarrow{(3)}$ CH$_2$NH$_2$

COOH   COOH

Cysteine   Taurine

Fig. 12.1 The formation of taurine from the amino acid cysteine. The enzyme catalyzing step (2) is deleted in cats, accounting for the dietary requirement of taurine in feline nutrition.

Canidae

Domesticated dogs, *Canis familiaris,* along with wolves, coyotes, foxes, and raccoons, belong to a family of primarily carnivorous terrestrial mammals, descended from wolf-like canines of the Pleistocene or glacial epoch about a million years ago. Again as in the case of the giant panda, the carnivore classification appears to be based on canine dentition, teeth adapted for crushing and cutting (the carnassials). Undoubtedly wolves, foxes, and raccoons are obligate (bona fide) carnivores. As for man's pets, the authoritative work (Hayes, 1985) lists requirements identical qualitatively with those of *Homo sapiens* and other mammals (rats, for instance), all of them omnivores. Presumably, therefore, the dog family, once domesticated, ceased to be obligate carnivores. However, canine dentition as well as the hunting instinct remained largely unchanged in most domestic species. Like foxes, our pets invade henhouses but they do not kill for nutritional reasons. Possibly they chew bones in order to supplement their meals with calcium or phosphorus.

Felidae

At least four genetic deficiencies are recognized and held responsible for the carnivorous lifestyle of the domestic cat, both the kitten and the adult. Most thoroughly studied (Hayes, 1988) is the cat's inability to produce taurine, a derivative of the amino acid cysteine, formed by a three-step process of which the second is deleted (Fig. 12.1).

Taurine, first isolated from ox bile (as the name indicates), is hardly a mainstream biochemical. Plants and bacteria do not need or produce it. Modern biochemistry texts barely mention it. Historically, taurine deficiency was an important discovery, the first example showing a causal relation between nutritional lifestyle and a genetic defect.

Until recently, taurine was known only in the form of taurocholic acid, a conjugate of cholic acid and taurine, known as a bile salt. In the bile of most vertebrate species, taurocholate occurs along with glycocholate, a cholic acid conjugate with glycine. Both bile salts are powerful detergents, serving to emulsify dietary fats prior to intestinal absorption. Cat bile, however, contains taurocholate exclusively.

Under normal circumstances, cats consume taurine-rich foods such as meat and fish. On synthetic diets lacking this molecule—which are adequate for rats and mice—cats, and especially kittens, develop a marked degeneration of the retina as measured by the electroretinogram. This test records changes in potential when an electrode is placed on the eye's surface. Ultimately an impairment of the entire photoreceptor population ensues, in kittens leading to blindness. Taurine levels are sharply reduced in blood plasma and most markedly in the brain, the organ richest in taurine. All these changes can be attributed to impaired cysteine decarboxylase activity; see Fig. 12.1, step (2). The role of taurine in the etiology of retinal degeneration remains to be defined, though it is known that scotopic—dim-light—vision in both rods and cones is severely affected. There is also mounting evidence for a neurotransmitter function in the cat's brain. Notably associated with retinal degeneration is the loss of visual acuity, the ability to resolve detail.

Only marginal effects of taurine deficiency are seen in newborn infants, especially when fed parenterally, and also certain monkeys (cebus) but not the Old World macaca species. At best, taurine deficiency in primates appears to result in less than optimal growth. So far at least, only the domestic cat requires taurine as an essential dietary constituent, one of several factors forcing this order to a truly or literally carnivorous lifestyle.

Finally, a molecule called felinine, an isopentenyl (branched five-carbon) derivative of the amino acid cysteine, is excreted uniquely in cat urine. This may be (my speculation) the consequence of the cat's genetic deficiency of the decarboxylating enzyme needed for the conversion of excess cysteine to taurine.

Vitamin A

The vast majority of animals satisfy their need for vitamin A by ingesting plants that contain the yellow-orange plant pigment β-carotene.* Containing thirty carbon atoms, β-carotene is not a vitamin per se. It has no physiological function in animals unless cleaved enzymatically into half-sized fragments, known as retinal (Fig. 12.2), which in turn reacts with the protein opsin to form rhodopsin (visual purple), the light absorbing pigment of the retina. The cat family lacking the β-carotene cleavage enzyme must therefore be provided with vitamin A itself, available only from other animal sources. (Cod liver oil, a rich source of vitamin A, is known, perhaps only to my generation, in the form of Scott's Emulsion, an unpleasant-tasting medicine.)

The symptoms of vitamin A deficiency in the cat include degenerative effects on the cornea as well as the retina, similar but not quite identical to those with deficiency syndromes seen in other species, including primates. Night blindness, loss of visual-light sensitivity, appears among the early symptoms, and total blindness in extreme deficiencies. Needless to say, for the cat, a hunter and predator, visual impairments are especially incapacitating and must be even more pronounced for the larger felines in the wild. A single such example is on record. It states that a captive African lion cub, presumably vitamin A–deficient, was afflicted with ataxia (lack of muscular coordination), night blindness, and anatomical changes in the brain.

*When I visited Chinese friends a number of years ago, I found them greatly concerned because of the bright red skin color of their infant, otherwise in perfect health. My inquiry about the youngster's diet elicited an admission that carrot juice was his favorite beverage. I had just read a *New Yorker* article by Berton Rouéchet describing similar cases of unusual skin pigmentation, which returned to normal on withdrawal of carrot juice. I suggested to my friends that they hold off on the carrot juice for a while. They did, and their baby's skin color promptly returned to normal.

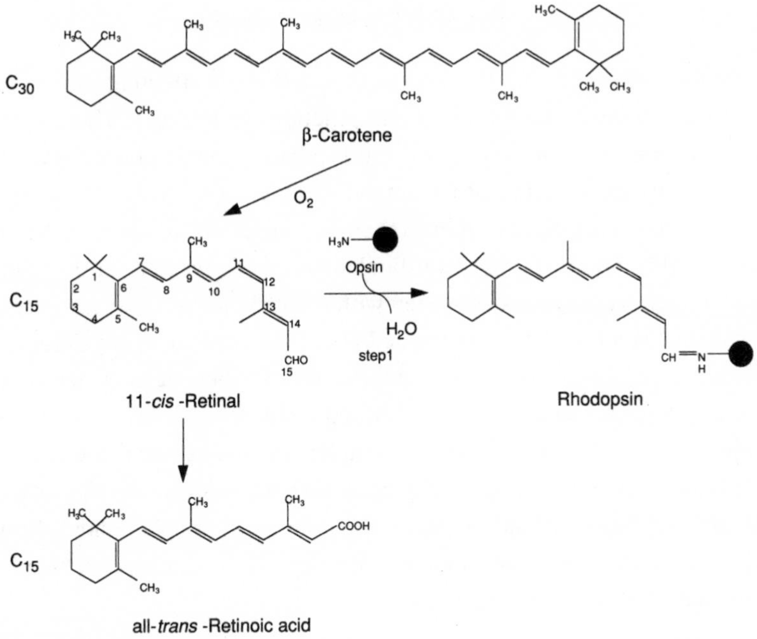

β-Carotene

O₂

C₃₀

11-*cis* -Retinal

Opsin

H₂O

step1

Rhodopsin

C₁₅

all-*trans* -Retinoic acid

C₁₅

Fig. 12.2 Potentially, β-carotene (C₃₀) could give rise to two molecules of retinal (C₁₅). Curiously, the stoichiometry of the cleavage reaction has not been determined. The black dot refers to the protein opsin.

The debilitating consequences of vitamin A deficiency have so far focused entirely on the degenerative changes in the visual apparatus. A second major role for vitamin A has now emerged, with the discovery that an oxidation product of carotene-derived retinol or retinal, called retinoic acid (Fig. 12.2), is concerned with cellular differentiation. Retinoic acid receptors in the cell nucleus have been recognized, leading to the hypothesis that retinoic acid may act much like a hormone, regulating protein synthesis at the level of transcription.* Whether retinoic acid formation from vitamin A is also crucial for differentiation in felines has not been studied.

*This would be another example of a vitamin that acquires hormonal properties only after ingestion. Thus, vitamin D is now known to regulate calcium and phosphorus absorption not per se, but only after some complex modifications in the body.

Not All Fat Is Created Equal

Before we proceed to the next genetic deficiency affecting felines, some remarks on fatty acid structure and function are in order. There is much current interest if not obsession, both popular and scientific, relating to the quality and quantity of fat optimal for human nutrition. Obesity, cardiovascular disease, and, more recently, cancer are believed to be associated with dietary fat. Several books have been written on the subject, and numerous conferences—often with the same set of speakers—are held annually. Much is still to be learned about the chemistry, physiology, and pathology of fats, and whether specific ones are beneficial or detrimental. As far as public health is concerned, common sense seems to be the way to go, at least for the time being. Complicating the debate is the enormous structural diversity of naturally occurring fatty acids, varying in chain length and degree of unsaturation (the number of double bonds). In what may be an extreme case, some five hundred fatty acid species have been isolated from the phytoflagellate *Euglena gracilis!*

On Fatty Acid Structure and Function

Recent guidelines from the Office of the Surgeon General and other federal bodies advise the American population to restrict the total fat content of their diets to 30 percent of the caloric intake. Palmitic acid (from palm oil) is to be minimized in the diet. This advice, emphasizing consumption of unsaturated fats, seems reasonable, but does not include adequate counsel about the kinds of fats.

All animals, yeasts, plants, and some bacteria produce oleic acid (unsaturated) from stearic acid (saturated), a process called desaturation and requiring oxygen. Desaturation refers to the abstraction of two adjacent hydrogen atoms from the fatty acid chain. A single desaturation produces oleic acid, a monounsaturated or monoenoic acid. Additional desaturations lead to dienoic acids, trienoic acids, and so on (Fig. 12.3). Oleic acid, in contrast to stearic acid, is a liquid at room temperature; olive oil is its most abundant source. In turn, oleic acid can be further desaturated

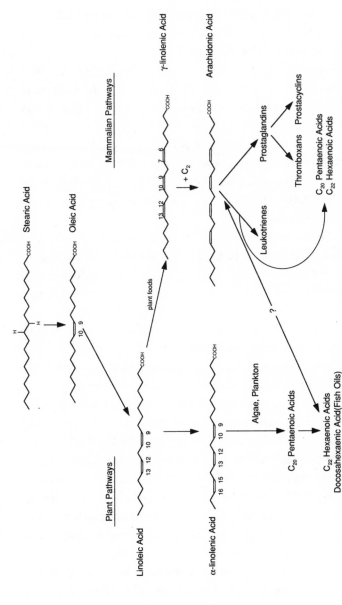

Fig. 12.3 Both animals and plants can convert stearic acid to oleic acid. Only plants convert oleic to linoleic acid, which is therefore essential for animals. Additional double bonds are introduced into linoleic acid in opposite directions, toward the carboxyl group in animals (ω-6) and toward the terminal methyl carbon in plants (ω-3). In both animals and plants, additional double bonds enter after chain elongation by two- or four-bond atoms.

to the dienoic linoleic acid, but only in plants. Animals therefore need an outside source of linoleic acid, essential for animal nutrition in the same way as vitamins and some amino acids are.

Essential fatty acid deficiency in rats was discovered by Burr and Burr in 1929. They described its symptoms, among them dry and scaly skin, retardation of growth, and both male and female infertility. In some species the symptoms develop slowly and become evident only in the offspring of deficient animals. Whether linoleic acid per se has a specific identifiable function is not known. It is needed, however, as a precursor of even more highly unsaturated or polyenoic acids, containing three, four, five, or six double bonds (Fig. 12.3), some of which are hormonal in character, acting locally as metabolic signals.

This pathway begins with the conversion of linoleic acid, the essential fatty acid animals derive from plant sources, to form the trienoic γ-linolenic acid, by oxygen-dependent insertion of an additional double bond in the direction of the carboxyl group. The trienoic acid, also lacking a recognized function as such, serves as an intermediate for the biosynthesis of the key essential fatty acid arachidonic acid, a polyunsaturated relative of arachidic acid, present in peanut (arachis) oil.

Arachidonic acid in turn undergoes a complex series of transformations leading to a family of bioactive molecules known as prostaglandins, so named because they occur in relatively high concentrations in prostate glands. Sheep prostates have proved to be the richest source. Crude extracts from thousands of sheep glands, supplied by the Upjohn Company in Kalamazoo, Michigan, enabled Sune Bergström at the Karolinska Institute in Stockholm to elucidate the chemical structures of the prostaglandin family (1962–1963). His colleague Bengt Samuellson identified the numerous steps leading from linoleic acid to the prostaglandins. Thus, three decades were to pass until the essential nature of linoleic acid in animal nutrition could be rationalized.

These momentous achievements led to many medical applications. Prostaglandin's diverse biological activities include stimulation of smooth muscle (uterine, for instance) and dilation of small bronchial arteries. To add to the complexity, prostaglandins are not the end points of arachidonic acid metabolism. They can be converted further to two major products: the prostacyclins and the thromboxans. Both have profound effects on blood platelet aggregation (blood clots), thromboxans causing aggregation and prostacyclins reversing it.* The foregoing discussion may be incomplete, explaining only one of the several consequences of linoleic acid deficiency affecting all animals, carnivores and herbivores alike.

This lengthy, yet all-too-superficial discussion is needed to introduce the reader to a further genetic disorder, the cat family's inability to generate arachidonic acid from γ-linolenate, *in addition to* the universal vertebrate deficiency of converting oleate to linolenate. A committee of the National Research Council has in fact recommended inclusion of both linoleate (18:2) and arachidonate (20:4) in synthetic diets deemed nutritionally adequate for cats. Some of the potential consequences of arachidonate deficiency for the cat are evident from the discussion above—failure to form prostaglandins and their bioactive derivatives.

For further, largely speculative rationalization of the feline's carnivorous lifestyle, we turn to some experiments with primate species (Neuringer, Anderson, and Connor, 1988). When rhesus monkeys were kept on a diet deficient in some polyunsaturated acids, a series called ω-3 (Fig. 12.3), dramatic changes occurred in the fatty acid composition of the retina, especially the content of docosahexaenoic acid, an acid containing twenty-two carbon atoms and six double bonds (22:6). Along with these compositional changes, profound abnormalities of the monkey's visual system were seen, notably a delay in the electroretinogram response of both retinal rods and cones, symptoms similar to those that cats experience in taurine deficiency. Furthermore, deficient animals showed a prolonged delay when recovering from a dark-adapted response after an initial flash of light. Thus in primates depletion of docosahexaenoic acid causes abnormalities in retinal photoreceptor function, including impaired visual acuity.

One wonders whether the essential fatty acid deficiencies, notably the requirement for arachidonic acid (20:4, ω-6), directly or indirectly also

---

*The key enzyme initiating prostaglandin synthesis from arachidonic acid, called cyclooxygenase, is inhibited by aspirin (acetyl salicylic acid), as demonstrated by John Vane in 1971. Thus, 120 years after the preparation of this analgesic, antipyretic, and anti-inflammatory drug, at least one mode of action could be defined in molecular terms. Salicylic acid, which is rapidly formed from aspirin in the body, is also an anti-inflammatory agent; but the biochemical basis is not clear, since it does not block cyclooxygenase.

affect the visual system of feline species. As noted above, retinal membranes in monkeys are rich in docosahexaenoic acid (DHA), and the same seems to be true of the domestic cat. What is the source of this docosahexaenoic acid for felines? Normally, either meat or plant or fish oils would provide it, but in deficient diets supplemented with arachidonic acid the source is not obvious. Pathways from arachidonic acid to DHA are not known and mechanistically improbable.*

Unfortunately, the experimental diets that revealed essential fatty acid deficiency in cats were not totally synthetic. The supplements were either a mixture of soybean oil and linseed oil (experimental diet) or, in the controls, a mixture of meat, fish, and fish-based cat foods, the latter containing docosahexaenoic acid. It is therefore not clear whether diets that caused essential fatty acid deficiency in cats did contain or lacked the fatty acid that normally appears in retinal membranes. I would guess that the cats' predilection for fish food meets their requirement for DHA, but this remains to be proven.

We have seen that both taurine and vitamin A deficiency impair retinal function of the cat. If the third deficiency, that of several essential fatty acids, should impair visual competence as well, one could argue that the carnivorous lifestyle of the cat seeks by every available means to repair or compensate for what would be a serious, possibly fatal disability in the feline's hunt for prey. To prevent retinal degradation associated with impaired visual acuity, including diminished night vision and, *in extremis,* blindness, constitutes the highest priority for the predator's survival. If correct, this hypothesis would alter our perception of obligate carnivores. Survival, not aggression or the lust to kill, is the driving force. The genes are at fault.

A fourth genetic deficiency of cats, their inability to form nicotinic acid from tryptophan, was mentioned in Chapter 3. It should therefore be possible to induce pellagra-like symptoms in feline species. Such experiments apparently have not been done.

---

*Some recent experiments with labeled linoleic acid fed to cats indicate a limited conversion to arachidonic acid. It is not yet clear whether the magnitude of this conversion is sufficient to meet the cat's requirement (Alexander Leaf and Norman Salem, private communication).

A thorough literature search has uncovered only one more instance of biochemically defined genetic defect in carnivores. Rivers and colleagues (1976) mention that the predatory turbot *Scopthalamus maximus,* a North Atlantic and Mediterranean flatfish, also lacks the enzyme for converting linoleate to linolenate and hence the route to arachidonate. One can infer that the rainbow trout and other fighting fish have the same dietary needs. Whether the consequences of these genetic defects in carnivorous fishes are similar to those established for the cat family remains to be seen. It is not unreasonable to postulate that dark vision and visual acuity generally are as essential for the survival of aquatic captors of prey as they are for terrestrial predators.

Raptors

Visual acuity has been stressed as one of the vital capabilities endowing obligate carnivores with the means for detecting their prey. Anyone who has witnessed raptors, the birds of prey, diving onto small rodents from great heights will be impressed by the remarkable precision of the prey-catching act. The term "visual acuity" refers to the eye's adaptation for recognizing objects at high resolution. Specialized central regions of the retina (pits or depressions named foveal receptors), containing cones but not rods, are responsible for the eye's resolving power. The greater the proximity of the foveal regions, the greater the eye's capacity for distinguishing an object's detail and the contrast between the object and its environment. Investigators studying visual acuity at the anatomical level have found inter alia that the raptor's foveal receptors are more closely spaced than they are in the human eye. Moreover luminance (light intensity) is a major determinant of visual acuity. So far at least, no information seems to exist for characterizing high and low visual acuity in biochemical terms.

Nevertheless, interesting differences in the modes of prey hunting by raptors have been revealed by the researches of Liz Raymond (1985, 1987). The wedge-tailed eagle *Aquila audax* achieves maximum acuity—twice that of man—at high luminances, consistent with the bird's lifestyle. The eagle hunts only in bright light (there is good reason for the proverbial "eagle eye") and roosts soon after sunset. By contrast, the visual acuity of *Falco berigora,* the Australian brown falcon, is only half that of the eagle but is relatively unaffected by diminished light intensity. As a consequence, this falcon successfully hunts not only in bright daylight, but also at dawn and dusk.

I had hoped to come across some biochemical or nutritional explanation for the carnivorous lifestyle of raptors comparable to the well-characterized genetic deficiencies of the domesticated cat. The only parallel, however, is behavioral: visual acuity to optimize success in the hunt for prey.

The Giant Panda

My original interest in carnivorous lifestyles was triggered by the unusual feeding habits of the giant panda. Why should this member of the order Carnivora pursue a lifestyle that is strictly herbivorous, thriving on a diet that consists solely of bamboo shoots? Little did I know that Carnivores are not necessarily carnivorous, but are classified by taxonomists on the basis of anatomical and morphological traits rather than nutritional needs. As mentioned, to the professionals specialized dentition is the landmark of Carnivora, and by this anatomical criterion alone the giant panda is assigned to the order. Perhaps this "plantigrade carnivore" needs these dental tools to crush, tear, and grind the tough bamboo plant. Still, why bamboo? Apparently the few animals in captivity are too precious for experimental purposes, for defining the panda's nutritional needs that only bamboo can satisfy.

Purely by chance, I came across a book entitled *Avanti il Panda!* by Richard Taylor, translated into Italian from the English edition, called *Next Panda, Please!* (Allen and Unwin, 1982). The story Taylor tells raises some doubts about the giant panda's nutritional classification and lifestyle.

A British veterinary surgeon, Taylor received an unusual request from the director of the zoo in Madrid. He was asked to travel to Spain in order to diagnose what ailed a giant panda pup named Chan-Chan. Taylor quickly established that Chan-Chan's affliction was a gastric ulcer. He placed the animal on a diet of milk, eggs, and honey, with the result that Chan-Chan rapidly gained weight. His ulcer healed within a few weeks.

After his recovery, a few bamboo shoots were added to the curative menu, but Chan-Chan showed no interest whatever in this supposedly obligatory panda food, and he never changed his mind.

The nutritional classification of this "obligate herbivorous carnivore" remains in limbo.

Bibliography

1. J. L. Gittleman (1989), *The carnivores: Carnivore behavior, ecology and evolution.* Ithaca, New York, Cornell University Press.

2. A. H. Brush (1990), Metabolism of carotenoid pigment in bones, *FASEB J.* **4,** 2969.

3. J. Tinoco et al. (1977), Docosahexaenoic acid in retinal lipids, *Biochim. Biophys. Acta* **486,** 575.

4. K. C. Hayes (1985), *Nutrient requirements of dogs.* Washington, D.C., National Academy Press.

5. K. C. Hayes (1988), Taurine nutrition, *Nutr. Res. Rev.* **1,** 99–113.

6. H. S. Geggel et al. (1987), Nutritional requirements for taurine in patients receiving long-term parenteral nutrition, *New Engl. J. Med.,* Jan. 17, 1987.

7. J. P. Rivers et al. (1976), The inability of the lion, *Panthero leo,* to desaturate linoleic acid, *FEBS Lett.* **67,** 269.

8. J. P. Rivers, A. J. Sinclair, and M. A. Crawford (1975), Inability of the cat to desaturate essential fatty acids, *Nature* **258,** 171.

9. M. Neuringer, G. J. Anderson, and W. E. Connor (1988), The essentiality of n-3 fatty acids for the development and function of the retina and brain, *Ann. Rev. Nutr.* **8,** 517–541.

10. S. Bergström (1982), The prostaglandins: From the laboratory to the clinic, *Les Prix Nobel,* Nobel Foundation, pp. 126–144. Stockholm, Imprimerie Royale, P.A. Norstedt.

11. B. Samuellson (1982), From studies of biochemical mechanisms to biological mediators, prostaglandin endoperoxides, thromboxanes and leukotrienes, *Les Prix Nobel,* Nobel Foundation, pp. 153–174. Stockholm, Imprimerie Royale, P.A. Norstedt.

12. L. Raymond (1985), Spatial visual acuity of the eagle, *Aquila audax, Vision Res.* **25,** 1477.

13. L. Raymond (1987), Spatial visual acuity of the falcon, *Falco berigora, Vision Res.* **27,** 1859.

13

Biochemistry's Origin and Future

——

When did biochemistry become a separate field of study? Historians will credit Lavoisier, Liebig, Pasteur, Bernard, and many other nineteenth-century scientists. They were ancestral figures coming from different disciplines but all interested in the chemistry of the life process. Professionally they included physicians, physiologists, biologists, botanists, chemists, and students of nutrition.

As early as 1878, Hoppe-Seyler founded the *Zeitschrift für Physiologische Chemie,* a journal that published papers on subjects such as the chemistry of blood and urine. Yet at the time no sponsoring society existed, nor was biochemistry as such a subject for teaching and research in academia. The universities of Tübingen and Liverpool were among the few exceptions.

Though opinions may differ, a good case can be made for Eduard Buchner as the person who (inadvertently) broke down the barriers—literally—and enabled biochemistry to become a science on its own. Buchner's very practical experiments in 1897 aimed at producing a palatable yeast preparation for nutritional purposes. At the time, whole yeast was a popular nostrum; as it turned out later, it was also a rich source of vitamins. Buchner, a chemist at the Agricultural College in Berlin, carried out a simple experiment. He ground yeast with sand and thereby disrupted the cells. He left the resulting aqueous suspension in a corked bottle overnight. Returning to his laboratory the next morning, he found the container uncorked. The foaming liquid had spilled all over the laboratory bench.

In essence, it was this unexpected outcome that earned Buchner the 1907 Nobel Prize in Chemistry, "for his biochemical researches in the discovery of cell-free yeast fermentation." Some chemists were puzzled. Adolph von Bayer, the dean of German organic chemistry, commented, "This will bring Buchner fame, even though he has no chemical talent."

Relevant or not to chemistry, Buchner's result would overturn Louis Pasteur's long-held and shared axiom that the chemical events of life are associated with and can be studied only in intact living cells. Typically, Buchner's experiment led to fortuitous results. Apart from the practical value of a potable and palatable yeast medicine, Buchner's infinitely more significant achievement was to prove that yeast fermentation takes place and can be studied in the soluble cell plasma, after the disruption of intact cells. This death knell for vitalism paved the way for identifying the various steps in alcoholic sugar fermentation and the enzymes that promote it. Yeast fermentation became the first process in what was subsequently known as intermediary metabolism. The reasons for choosing yeast were practical and commercial, not scientific. At any rate, yeast became as important an organism for early biochemistry as did the bacterium *Escherichia coli* for the beginnings of molecular biology.

Other studies that furthered the science of biochemistry in the United States were carried out at the Connecticut Agricultural Station and the Sheffield School at Yale. R. H. Chittenden, T. B. Osborne, and L. B. Mendel, examining the proteins of plants, determined that some cereal grains were of higher nutritional quality than others. Differences in the amino acid composition of plant and animal proteins were found to be responsible. Eventually amino acids could be classified either as nonessential (dispensable) nutrients, or as essential protein constituents. The animal body synthesizes the former, while relying on an external supply of the latter.

William C. Rose, trained at Yale and later a member of the chemistry department at the University of Illinois, deserves the credit for establishing and identifying the two classes of amino acids in human nutrition. Graduate students in chemistry were Rose's experimental subjects. Each individual, for a period of several weeks, followed a synthetic diet lacking one of the essential amino acids. A positive nitrogen balance (that is, urinary excretion of excess nitrogen) indicated that the missing amino acid was essential for protein synthesis and, as learned later, for various metabolic functions. The University of Illinois was probably the first establishment to recognize biochemistry as a respectable branch of chemistry for teaching and research.

Equally prominent among the midwestern institutions that began as agricultural or land grant colleges and turned to biochemistry was the University of Wisconsin at Madison. It became an early center of vitamin research, responsible for the discovery of two essential animal growth factors: the "sunshine" vitamin, D (an antirachitic), and niacin (the anti-pellagra factor). This biochemical research was carried out in the Department of Agricultural Biochemistry. E. B. Hart, C. A. Elvejhem, and H. Steenbock were distinguished members. The discovery of vitamin D formation by ultraviolet radiation of yeast sterols was to bring substantial benefits to the university. A lucrative patent was issued to the Wisconsin Alumni Foundation and is to this day a major source of research funds for the university. The patent was controversial, probably the first granted for a natural product. The issue plagues legislators even now, notably in the biotechnology industry.

This all-too-brief account neglects the many contributions other fields such as microbiology, physiology, and medicine have made to the beginnings of biochemistry as a separate discipline. Here I admit to a bias in favor of American institutions, but the contributions in Great Britain, France, and Germany were of equal importance. For detailed information the reader should turn to the masterly historical account of the field by Joseph S. Fruton (*Molecules and Life,* New York, John Wiley, 1972).

Biochemistry and Molecular Biology

This writer learned his biochemistry from a text written by Oskar Bodansky in the mid-1930s. In some three hundred pages it contained all the essential information a beginning biochemist needed. By contrast, up-to-date texts today approach the size and weight of a Webster's unabridged dictionary. Not only has traditional biochemical information expanded exponentially, but under the same umbrella the revolutionary developments of molecular biology and much of cell biology and genetics are treated as well.

I do not know who first named this new science "molecular biology"; for it is inseparable from modern biochemistry. Most likely, one of the numerous physicists who turned to biology without background or interest in chemistry was responsible for the baptism. I remember an episode that supports this explanation. When James Watson joined the biochemistry department at Harvard University in the 1960s, he expressed discomfort at belonging to a department so named, claiming that his minimal knowledge of chemistry disqualified him from membership.*

A growing number of biochemistry departments have recently changed their names to "biochemistry and molecular biology," a tautology justified only on historical grounds. All chemicals, biological as well as man-made, are molecular. But not all biology is chemical. Designations such as "chemical biology" would be more truly descriptive and make more sense. Perhaps it is a trivial issue. In any event, no terminology is forever.

Ingenuity and Intuition

In the mid-nineteenth century, the French philosopher Auguste Comte articulated what he called a hierarchy of the natural sciences. His doctrine ordered various sciences chronologically (mathematics, physics, chemistry, and biology), gave their objects and their objectives, and ranked their

---

*In Chapter 26 of his classic, *The Double Helix* (New York, Atheneum Press, 1968), Jim relates an event of momentous consequence during his and Francis Crick's building of a DNA model compatible with the x-ray diffraction data. The chemist Jerry Donohue was their neighbor in the Cambridge laboratory, at a time when the mode of base pairing in DNA had not yet been settled. Donohue, inspecting a tentative model, pointed out that Watson and Crick were probably in error to write the structures of the bases guanine and thymine in the "enol" (R-C-OH) rather than the tautomeric "keto" (RH-C=O) form. Jim's defense: "The enol tautomer I have copied out of Davidson's book and many other texts of the time." In Donohue's opinion, all the texts were wrong, as indeed they were—or, more likely, were not up to date. Written in the keto form, the nucleic bases paired "correctly." Thus a temporary logjam in DNA model building was broken. A chemist's intuition had played a critical role in the structural elucidation of biology's most celebrated molecule.

inherent degree of reality. Biochemistry, a relative latecomer as a scientific discipline, Comte could not list in his hierarchy. Biology and chemistry, previously separate disciplines, tended to overlap increasingly as both sciences extended their frontiers. Beginning at the turn of the twentieth century, this offspring became a hybrid of exceptional vigor.

As for the depths of insight this new science was able to provide, it shared biology's limitations on generalizing or predicting phenomena. There are no laws of biology comparable to those of physics—thermodynamics, gravity, and the like. (Even the "laws of Mendelian inheritance" have now been downgraded to "Mendel's ratios of inheritance.") Both in biology and in biochemistry, hypotheses, axioms, postulates, and sometimes dogmas are code words, all tending to have limited life spans.

In "pure" chemistry, in contrast to biochemistry, many reactions are governed by rules; but rules allow for exceptions ("as a rule"). A notorious example is the exception to Blanc's rule, which delayed identification of the structural identity of the cholesterol molecule for nearly a decade. The formation of certain sterol degradation products failed to behave as predicted. X-ray crystallography provided the correct answer.

During most of my career, both chemists and biochemists have been my colleagues. How different their approach to a given experimental problem! The chemist, in whatever field of specialization, is in charge, a commander having at his or her disposal numerous variables—temperature, atmospheric pressure, assorted catalysts, and usually a choice of reactants. Depending on the experimenter's skill and ingenuity and a vast reservoir of known precedents, the chemist will ultimately arrive at the specified goal, no matter how complex. Very often this goal will be realized by several routes, some more elegant and some more cumbersome. In organic synthesis, "elegant" indeed constitutes the highest praise. To the biochemist aiming at the same goal, Nature, not the experimenter, prescribes the experimental conditions. They are fixed, a given in the living cell.

Let me quote from François Jacob's autobiography (*The Statue Within,* 1988): "Starting with a certain conception of the system, one designs an experiment to test one or another aspect of this conception. Depending on the results one modifies the conception to design another experiment— and so forth. That is how biology works." One may add, And so does biochemistry.

To mention cholesterol once again, the chemical synthesis of this molecule became a realistic if formidable enterprise once its structure was firmly established. R. B. Woodward and D. Barton achieved this goal elegantly in 1951, aided by the wealth of knowledge at their command. Needless to say, their design did not and could not imitate Nature's still totally unknown design for accomplishing the same feat. Only decades later was the biological route for cholesterol biosynthesis untangled and rationalized post factum. In this effort the chemical synthesis provided no guide whatsoever.

Some chemists foresee that their elegant synthetic devices, based on blueprints designed at the desk, will come close to if not equal Nature's inventions for producing complex molecules such as morphine and other alkaloids, antibiotics, and so on. Their optimism is understandable, but requires a vastly greater knowledge of Nature's logic than we possess today.

A young scientist facing the decision whether to opt for a career in a branch of pure chemistry or to explore Nature's way of making a molecule of life will have to asses his or her talents. The individual will have to decide whether to be a commander, exploiting the arsenal of available chemical compounds in ways limited only by the experimenter's ingenuity, or whether to let Nature be in charge and be an observer, François Jacob's "tinkerer," who induces biological systems to reveal their secrets by trial and error. Of course, such choices need not be irrevocable. Career changes of young or even middle-aged scientists are not rare. Increasingly, scientists trained in pure chemistry are turning to a branch that would rank lower in an updated version of Comte's hierarchy; to biological chemistry—and even beyond, to molecular biology and genetics. (I must confess that I was born too early to be converted.) Notably, moves in the opposite direction seem to be rare. Have there been trained biologists who later made a mark in any branch of pure chemistry? This phenomenon is perhaps not as strange as it might seem, given that there are no formal barriers. In all probability it is primarily a language problem, not an intellectual problem. For chemistry, more than any other scientific discipline, requires learning and memorizing a foreign language and a vast vocabulary—a task that becomes more arduous with advancing age.

Glossary

———

Alkaloids: Organic plant substances, usually containing a nitrogenous ring. Morphine, cocaine, and strychnine are examples.

Amino Acids: General structure

$$\underset{\underset{NH_2}{\mid}}{\overset{\overset{H}{\mid}}{R-C-COOH}}$$

They are the building blocks of proteins, differing in the nature of the R residues. Nine of the twenty amino acid constituents of protein are essential, but not made in the animal body. In aromatic amino acids, R is a derivative of benzene, indole, or imidazole. In aliphatic amino acids, R is a hydrocarbon chain.

Antagonists: Molecules that block the positive action of *agonists* such as drugs or metabolic signals.

Antimetabolites: Usually drugs that counteract the activity of essential metabolites, for example, vitamins or coenzymes.

ATP: Shorthand for adenosine triphosphate, which contains the nucleic base adenine linked to the sugar ribose, which in turn carries a triphosphate group AMP—P—P. Oxidation of fuels provides the energy to produce ATP, which then serves as the energy source for most biosynthetic reactions. This potential energy is made available by breaking the AMP—OP—OP bond to form AMP + POP. The ATP cleavage also converts chemical energy into mechanical work such as muscle contraction.

Bile Acids: Produced in the liver by oxidation of cholesterol and converted to bile salts by combination with the amino acid glycine or the amine taurine. Bile salts are detergents that aid in the dispersion and intestinal absorption of fats.

Chemical Bonds: The various linkages that hold atoms together by sharing one or more pairs of electrons. (1) *Strong* or covalent bonds: C—C, single bonds; C=C, double bonds (olefinic), and triple (acetylenic) C≡C, bonds.

(2) *Weak* bonds: electrostatic, for example, Na^+Cl^-; hydrogen bonds such as $C=O\cdots NH—$, which form and break in water much faster than covalent bonds; dominant in the interactions between peptide chains in proteins and between complementary nucleic acid base pairs, responsible for the double-helix structure of DNA. (3) *Hydrophobic interaction:* attractive forces between adjacent hydrocarbon structures, for instance, fatty acids; also known as van der Waals interactions.

Chromatography: The technique for separating and purifying molecules based on their differential adsorption on charcoal, alumina, or various synthetic polymers. The botanist Michael Twsett first applied it to the separation of plant pigments.

Coenzymes: Heat-stable, dialysable molecules that function only in concert with enzyme proteins. Usually they are the catalytically active regions within an enzyme-coenzyme complex. In a few instances, coenzymes are attached covalently to substrate molecules such as fatty acyl—CoA, or carboxybiotin, the enzyme bound active form of Carbon dioxide.

Dismutation: Molecular rearrangement, one group of the molecule being oxidized and the other reduced.

Endoperoxides: Internal —O—O— bridged structures within a six-membered carbon ring, as in some terpenes; potential sources of hydrogen peroxide.

Endophyte: A plant living within another plant.

Epimers: Molecules of identical elementary composition differing in physical properties because of the opposite spatial orientation of substituents at single carbon atoms.

Etiolatel Leaves: Leaves of higher plants that have been blanched or bleached through loss of chlorophyll.

Eukaryotes: See **Kingdoms.**

Fatty Acids: (1) Saturated: $CH_3(CH_2)_n COOH$; stearic acid ($n = 16$) and palmitic acid ($n = 14$) are the most abundant; (2) Unsaturated or olefinic acids contain one or more double bonds (—C=C—). Fatty acids occur predominantly linked to glycerol, the triglycerides or neutral fat found in plasma and in fat cells or adipocytes, the fat depots. See also **Phospholipids.**

Feedback control: A biosynthetic end product inhibiting an early reaction in a multistep pathway; for example, cholesterol inhibits the rate of conversion of hydroxymethyl-glutaryl CoA to the next product in the pathway leading to cholesterol.

Ganglions: A mass of tissues composed of nerve cells, in gray matter and in the spinal cord.

Gene Expression: Genetic information; the genetic code stored in DNA instructs cells to produce specific proteins.

Glycogenic: Amino acids that can be converted metabolically to glucose or glycogen.

Helminths: Parasitic flatworms.

Heterotrophic Organisms: Those that obtain nourishment from the outside; by contrast, autotrophic organisms synthesize all needed organic compounds within their cells.

Hormones: Chemicals produced and secreted into the bloodstream by an endocrine gland (such as the thyroid or pancreas), producing a specific biological response in a remote organ or tissue. Hormones are messenger molecules.

Isomers: Molecules of identical elementary composition and molecular weight, but differing in the spatial arrangement of atoms in the molecule. Stereoisomers or enantiomorphs refer to compounds having four different substituents linked to the same asymmetric carbon atom, such as the α-carbon of amino acids.

Isotopes: One or more species of the same element having the same atomic number, identical in chemical behavior but of different atomic mass. Isotopes are of two kinds, stable and radioactive. Important for research in chemistry and biology are the hydrogen ($^1H_2$) isotopes, $^2H_2$ or deuterium (stable) and $^3H_2$ or tritium (radioactive); and the isotopes of carbon ^{12}C, the stable ^{13}C and the radioactive ^{14}C. Isotopes of oxygen, phosphorus, and nitrogen also exist and are widely used in a methodology known as the tracer technique.

Ketogenic: Amino acids that can be metabolized to the ketone bodies acetone and acetoacetic acid, which are excreted in the urine in severe diabetes.

Kingdoms: The highest category in the hierarchy of organismic classification. The three kingdoms are (1) single-celled or monera prokaryotes, including eubacteria and archaebacteria; (2) protista, unicellular eukaryotes, subkingdom of animals; (3) multicellular higher plants and animals. Prokaryotes, the earliest cells, lack a membrane-bound nucleus or other organelles. Eukaryotes are the most highly developed multicellular organisms containing subcellular organelles (for example, a membrane-bound nucleus, mitochrondria, the endoplasmic reticulum, and several others).

Laser Radiation: Light amplification stimulated by emission of radiation.

Ligands: Molecules that bind to a target, specifically of signals to receptors.

Membranes: Define the boundary between the inside and outside of cells or intracellular organelles. Molecules entering or leaving cells must pass through a membrane barrier.

Metaphyta: Multicellular plants.

Metazoans: Multicellular animals.

Mevalonic Acid (MVA): A six-carbon precursor of cholesterol, formed from three molecules of acetic acid. The cellular level of MVA controls the synthesis of cholesterol. Synthetic analogs of MVA are effective drugs that reduce the levels of blood cholesterol and of the plasma lipoprotein LDL.

Monomer: Low-molecular subunits which can aggregate to larger entities—dimers, trimers, and ultimately polymers. In such aggregates the subunits may either be linked covalently (as glucose in glycogen) or noncovalently by hydrogen or hydrophobic bonds in many proteins (hemoglobin, for instance, consists of four monomeric polypetide chains).

Neurotransmitters: Chemical signals released by electrical signals in the brain that stimulate and act on targeted adjacent cells; short-range chemical mediators such as acetylcholine, dopamine, serotonin, and enkephalins. All bind to specific receptors.

Nucleotides: Precursors of DNA and RNA, containing the nitrogenous DNA bases adenine, guanine, cytosine, and thymine—and uracil instead of thymine in RNA. The nucleotide bases are linked to ribose in RNA and to the five-carbon sugar deoxyribose in DNA. The individual base-ribose or base-deoxyribose entities are linked by phosphate bridges in both RNA and DNA.

Oxidative Phosphorylation: A vectorial process for generating energy in the form of ATP coupled to electron transfer between cytochromes, and ultimately to oxygen.

Oxygenases: Enzymes that catalyze the entry of oxygen into organic compounds. Oxygenases are specialized cytochromes, hemoproteins named P_{450} enzymes, which must be distinguished from oxydases that catalyze the transfer of electrons from reduced metabolites to oxygen.

Peptides: A combination of two or more amino acids linking the carboxyl group of one to the amino group of another:

$$R_1COOH + NH_2R_2 \rightarrow R_1CONHR_2$$

Phospholipids: Water-insoluble constituents of membranes or cell envelopes, consisting of fatty acids, phosphorus glycerol, and nitrogenous bases such as choline or ethanolamine.

Polypeptides: Combinations containing up to fifty amino acids; when this number is exceeded, the molecules are called proteins, insulin composed of fifty-one amino acid residues being the smallest.

Polyproteins: Large biologically inactive proteins convertible by peptide bond cleavage to one or more biologically active messenger molecules; for example, preproinsulin → proinsulin → insulin, and preopiocorticotropin → corticotropin, melanocyte-stimulating hormone, endorphins, and enkephalins.

Prokaryotes: See *Kingdoms*.

Protein Functions: Several can be distinguished: (1) enzyme catalysis, that is, proteins accelerate the rate of biochemical reactions; (2) protein hormones act as metabolic signals; (3) molecules that carry small metabolites in blood and lymph; (4) chaperones control the three-dimensional structure of other proteins; (5) structural proteins such as collagen; (6) coproteins determine or alter the specificity of enzymes.

Protozoans: Free-living single-celled eukaryotes. Also known as "first animals."

Racemization: A chemical or biological process that changes D (dextro) to L (levo) or vice versa. Compounds such as the natural amino acids contain asymmetric carbon atoms. They are optically active, rotating the plane of polarized light when examined in a polarimeter. They exist in right-handed (D) and left-handed (L) forms, which are mirror images, not superimposable. All natural amino acids have the L-configuration, with the few exceptions noted in the text.

Radicals: Commonly called "free" radicals, formed by removal of one electron from an electron pair. A single dot, RO · indicates a free radical. Oxygen free radicals are highly toxic, attacking DNA and unsaturated fatty acids, among other substances.

Receptors: Membrane-associated proteins having a specific binding affinity for hormones, drugs, viruses, and toxins.

Ribosomes: Organelles consisting of several forms of RNA and numerous proteins. The synthesis of proteins from activated amino acids takes place in the ribosomal machinery.

Ruminants: Strict herbivores that possess a four-chambered stomach and chew the cud.

Tautomers: Forms of the nucleic bases that result when hydrogen atoms change their location, for instance, $NH_2 \rightarrow N = H$ or $C = O \rightarrow C - OH$. Important for the "correct" base pairing in DNA's double helix.

Teratogenic: Causing developmental malformations or monstrosities.

Terpenes (derived from turpentine): Volatile constituents of higher plants, especially conifers. Monoterpenes contain a single six-membered ring and various polyterpenes contain up to five fused rings.

Index